Teach Your Kids Money

A Parent's Guide to Raising Financially Capable Adults

Joseph Trek

Published by NobleTrex Press

For permissions and other inquiries, write to:

P.O. Box 3132, Framingham, MA 01701,
USA

Contents

Introduction

Imagine for a moment that you decided to teach your child to drive. But instead of starting in an empty parking lot on a Sunday morning, explaining the brake and the gas, and logging hours of practice with you in the passenger seat, you took a different approach. You drove them around for eighteen years, never explaining what you were doing or why. You hid the dashboard so they couldn't see the speedometer. If they asked how the car worked, you hushed them and said it was rude to talk about mechanics. Then, on their eighteenth birthday, you tossed them the keys to a Ferrari, pointed toward the highway, and said, "Good luck! Try not to crash."

It sounds insane. It would be negligent. It would be dangerous.

And yet, this is precisely how most of us handle money with our children.

We treat our financial lives as classified information. We shield our kids from the reality of bills, the logic of trade-offs, and the mechanics of earning, hoping to protect their innocence. We assume they will figure it out eventually, perhaps through osmosis or a high school economics elective. Then, young adulthood arrives. Suddenly, they are expected to navigate a world of predatory student loans, invisible digital payments, credit scores, and retirement planning-all without ever having sat in the driver's seat.

Is it any wonder that so many young adults feel paralyzed by anxiety when they check their bank balance? Is it surprising that credit card debt has become a rite of passage?

We can do better. And the good news is, it doesn't require you to be a math whiz or a wealthy tycoon. In fact, if you've ever made a financial mistake in your life, you are perfectly qualified to teach this subject. Your experience-scars and all-is the textbook.

This book is your manual for the driving lessons. It is about moving your child from the passenger seat, where they passively consume, to the driver's seat, where they actively manage resources. It is about raising adults who don't just survive the modern economy but possess the agency to shape their own lives.

The Invisible Curriculum

Here is the unsettling truth: You are already teaching your kids about money. You are teaching them every single day.

When you stress-eat takeout because the budget is tight, you are teaching them about emotional coping mechanisms. When you argue with your partner about spending in hushed, angry tones behind a closed door, you are teaching them that money is a source of conflict and shame. When you tap your phone to pay for groceries without a word, you are teaching them that resources are infinite and friction-free.

Children are observational geniuses. They are constantly downloading the "invisible curriculum" of your home. The question isn't *if* they will learn about money from you; it's *what* they will learn.

For generations, the default strategy for parents was silence. Money was considered private, unseemly, or strictly "adult business." But in the analog era, kids could at least see the mechanics. They saw you counting cash at the register. They saw you writing a physical check and balancing a checkbook. They saw the tangible exchange of labor for goods.

Today, money has become a ghost. It is a digital abstraction. Earning is often an invisible deposit; spending is a double-click on a screen; debt is a number in a cloud. For a digital native, the link between work, money, and stuff is broken. To them, Amazon packages appear on the porch like magic. Without intentional intervention, a child's mental model of money will be fundamentally flawed. They will see the consumption, but they will miss the calculation.

We have to make the invisible visible. We have to narrate the process. We have to break the taboo.

The Sandbox Strategy

If you take only one philosophy from these pages, let it be this: The stakes will never be lower than they are right now.

Many parents hesitate to give their children financial autonomy because they are afraid the child will mess up. They worry the child will waste their allowance on junk, lose their debit card, or forget to save.

Let me be clear: You *want* them to mess up.

You want them to blow their entire month's allowance on a plastic toy that breaks in twenty minutes. You want them to feel the hot sting of buyer's remorse when they are eight years old and the loss is ten dollars. Why? Because the brain learns from consequences, not lectures. That ten-dollar mistake is a tuition payment for a lesson in quality and impulse control.

If you protect them from small failures now, you are setting them up for catastrophic failures later. It is infinitely better for your child to bounce a check (or decline a transaction) when they are fifteen and living in your spare room than to default on a mortgage when they are thirty-five with a family of their own.

Throughout this guide, we will focus on building a "sandbox"- a safe, controlled environment where your kids can experiment with money, make terrible decisions, suffer the natural consequences, and try again, all without ruining their credit or their future. We are going to calibrate the stakes so they can practice the art of decision-making while the safety net is still firmly in place.

A New Definition of Wealth

Before we dive into the tactics of allowances and index funds, we need to redefine the goal. What does it mean to be a "financially capable adult"?

It doesn't necessarily mean they will be rich. We cannot guarantee our children will be millionaires, and frankly, that's a shallow target. Instead, we are aiming for something more robust: capability.

A financially capable adult understands that money is a tool, not a master. They can distinguish between a need and a want in a culture that screams everything is essential. They understand the difference between looking wealthy and being wealthy. They possess the resilience to handle an emergency without spiraling into panic. Most importantly, they understand that money is a finite resource that must align with their values.

This brings us to the heart of the matter. Money is rarely just about math. If it were, everyone with a calculator would be wealthy. Money is about psychology, habits, and values.

That is why we won't start this journey with spreadsheets. We will start with culture. We will begin by digging into what your family actually values, how to define your "enough," and how to talk about trade-offs without making money feel like a burden. We have to build the emotional foundation before we erect the structural walls of budgeting.

The Journey Ahead

So, how do we turn these high-minded ideals into Tuesday afternoon realities?

We will move through a progression that mirrors your child's development, shifting from simple concepts to complex systems. We begin by establishing the environment-the family money culture. You'll learn how to answer the awkward questions ("Are we rich?", "How much money do you make?") without lying or oversharing. We'll discuss how to neutralize the shame often associated with money so your children feel safe coming to you with problems later.

From there, we have to instill the core mental models. Before a child ever holds a dollar, they need to understand the concept of trade-offs. We live in a world of infinite options but finite resources. Teaching a child that "money is choices" is perhaps the most critical lesson of all. We will explore how to teach the difference between needs and wants, not as a lecture, but as a daily practice of prioritization.

Then, we get practical. We will dismantle the "allowance" debate. Should you pay for chores? Should allowance be free? The answer depends on what you are trying to teach, and we will look at the models so you can design one that fits your family. We will look at the power of "buckets"-visualizing savings so that a child isn't just saving for "the rainy day" (which sounds boring) but for the Lego Death Star (which sounds amazing).

As your children age, the challenges become external. We will tackle the consumer machine. Marketing departments spend billions of dollars employing psychologists to hack your child's dopamine receptors. We will equip your kids with the shields they need: impulse control strategies like the "24-hour rule," the ability to decode advertising, and the wisdom to resist the comparison trap of social media.

We will then navigate the digital plumbing of modern finance. The days of piggy banks are fading; your teenager needs to understand banking apps, peer-to-peer payments, and cybersecurity. We will cover the nuts and bolts of protecting their identity and their funds in a cashless world where scams are evolving faster than software updates.

As we move toward the teen years, we face the double-edged sword: credit and debt. These are the power tools of finance-useful for building a life, but capable of cutting off a limb if mishandled. We will demystify credit scores, explain the terrifying math of compounding debt, and have the hard conversations about student loans and the return on investment of higher education.

Finally, we will look to the horizon: investing and independence. We will demystify the stock market, stripping away the jargon of Wall Street to reveal the simple, boring, effective truth of long-term growth. We will show how time is your child's greatest asset, and how a few hundred dollars invested now can do the heavy lifting for their future selves. And we will prepare a launchpad for that final transition-handling the lease, the insurance, the taxes, and the grocery bill-so that when they walk out your door, they are looking forward, not looking back for a bailout.

Your Role as the Guide

You might be reading this and thinking, *But I'm terrible with money. I have debt. I don't have a retirement plan. Who am I to teach this?*

Let me relieve you of that burden right now. You do not need to be perfect to be effective. In fact, your vulnerability is an asset.

If you are digging yourself out of debt, take your teenager along for the journey (in an age-appropriate way). Let them see you making the hard choices. Let them see you cutting back on discretionary spending to attack a balance. Let them see the grit required to change a financial trajectory. That lesson-the lesson of resilience and course correction-is more valuable than any inheritance you could leave them.

Financial literacy is not a gene. It is not an innate talent. It is a set of learned skills and habits. It is a language. And just like any language, it is best learned through immersion, conversation, and practice.

The world is going to try to teach your kids about money. Credit card companies, luxury brands, influencers, and predatory lenders are all lining up to offer their curriculum. Their lesson plan is designed to extract wealth from your child. Your lesson plan is designed to build it.

By holding this book, you have already made the choice to intervene. You are choosing to hand over the keys, but only after you've taught them how to drive.

Let's get started.

Chapter 1

Building a Family Money Culture

*Money is often a source of silent tension in families, but it doesn't have to be. Before we talk about allowances or bank accounts, we need to talk about *us*-what our family values, how we speak about trade-offs, and how we handle mistakes. This chapter is your blueprint for creating a home where money isn't a taboo subject, but a tool for living a life that aligns with your beliefs.*

1.1 Defining Your Family's Money Values

Saturday morning, big-box store, fluorescent lighting. Your kid is holding a $39.99 remote-control dinosaur like it's a sacred artifact. You feel the familiar pressure: the other parent who seems to say yes with ease; the part of you that remembers being told no; the part of you that thinks, *If I don't buy this, am I being stingy?* You can do the usual dance—"Maybe later," "It's too expensive," "Put it on your birthday list." Or you can do something quietly powerful: treat the moment as a values decision, not a price decision.

That shift matters because most family spending conflicts aren't actually about the money. They're about what the money is *for*. Kids don't naturally understand that a family can have enough income and still say no. They also don't understand that a family can be under financial strain and still say yes to certain things on purpose. Without values, every request becomes a negotiation. With values, requests become a chance to practice a skill: matching choices to priorities.

A "money value" is a belief that shapes how you use money. Not what you *like* in a vacuum ("We like vacations"), but what you're willing to *trade for*. Values show up when you have to choose. If you value security, you keep a cash cushion even when the sales are tempting. If you value learning, you'll pay for museum memberships and tutoring while driving an older car. If you value generosity, you plan giving the same way you plan groceries. The point isn't to pick the "right" values. The point is to name *your* values so you can stop living by default settings—advertising's values, your childhood's values, your friends' values, or the values of the loudest kid in the aisle.

Most families already have money values. They're just unspoken, like the rules of a game you were never handed. Unspoken values create confusing whiplash: "We don't waste money...but we order delivery three times a week." "We're frugal...but we upgrade phones every year." "We care about experiences...but we're always too busy to take them." Kids are excellent pattern detectors. If what you say doesn't match what you do, they won't conclude you're hypocrites. They'll conclude money is mysterious, emotional, and unpredictable—something you argue about behind closed doors and then spend impulsively in public.

A useful way to think about values is as a translation device. Life hands you endless options; values translate them into *yes*, *no*, and *not right now*. Values are also a permission slip. When you know what you're prioritizing, you can decline the rest without guilt. Guilt is expensive. It turns a "no" into a resentful "fine," or a "yes" into a secret credit card balance. Clear values let you buy things with clean hands.

Here's the uncomfortable truth: if you don't decide what matters, your spending will still be perfectly aligned to *something*—stress relief, convenience, status, boredom, nostalgia, fear. Those are values too; they're just the kind that don't like daylight. One of the healthiest things you can do for your kids is to show them that adults make trade-offs on purpose, out loud, without shame. They learn that money is not a morality test. It's a tool for building a life.

From ``needs vs. wants" to ``wants vs. values"

Most parents start with the classic lesson: needs versus wants. It's helpful, but it breaks down faster than you think. A winter coat is a need. A $180 brand-name winter coat is...complicated. A phone might be a want for a 10-year-old and a need for a 17-year-old with a job and after-school activities. And then there are the things that aren't needs but still matter deeply: music lessons, a family trip, a summer camp where your shy kid finds their people.

A more grown-up framework is *wants versus values*. Wants are immediate; values are durable. Wants say, "I want it." Values say, "I want that kind of life." Kids can learn this earlier than we give them credit for. You can say, "We want a lot of things. We can't do all of them. Our family chooses to spend more on *X* because *X* matters to us."

Notice what that sentence does. It doesn't turn money into a scarcity monster ("We can't afford anything"). It also doesn't turn money into a free-for-all ("Sure, why not?"). It turns money into a planning problem: *We choose.*

The ``Shared Priorities" exercise (that doesn't feel like a board meeting)

If the phrase "family values discussion" makes you picture a sad circle of folding chairs, good. Don't do that. Do it like you'd plan a trip or a birthday party: with energy, specifics, and snacks. Your goal is to produce a short list of "Shared Priorities"—usually three to five—that describe what your family is intentionally building.

Start with a simple prompt: *When money is working well in our family, what does it make possible?* Let answers be messy. One kid might say, "Ice cream." Another might say, "A dog." You might say, "Less stress." Your partner might say, "More freedom." These aren't wrong; they're raw materials.

Then help everyone level up from objects to outcomes:

- "Ice cream" might become "small treats without guilt."
- "A dog" might become "a home where we can care for something together."
- "Less stress" might become "a buffer so surprises don't break us."
- "More freedom" might become "choices: time off, travel, changing jobs."

Next, turn outcomes into priorities that can guide real decisions. Examples:

- *Security*: we keep a cushion and avoid money emergencies.
- *Experiences*: we choose memories over upgrades.
- *Health*: we spend to support sleep, movement, and good food.
- *Learning*: we invest in skills, books, classes, and curiosity.
- *Generosity*: we give in a planned way, not only when we feel guilty.

Finally, add the missing ingredient: *what we're willing to give up*. This is where values become real. "We value experiences" means you might skip the premium cable package. "We value security" means you might drive the paid-off car longer. "We value health" might mean fewer impulse snacks at the checkout line and more money toward sports fees or cooking tools.

If you want your kids to respect the priorities, let them see the trade-offs. Values that never cost anything aren't values; they're slogans.

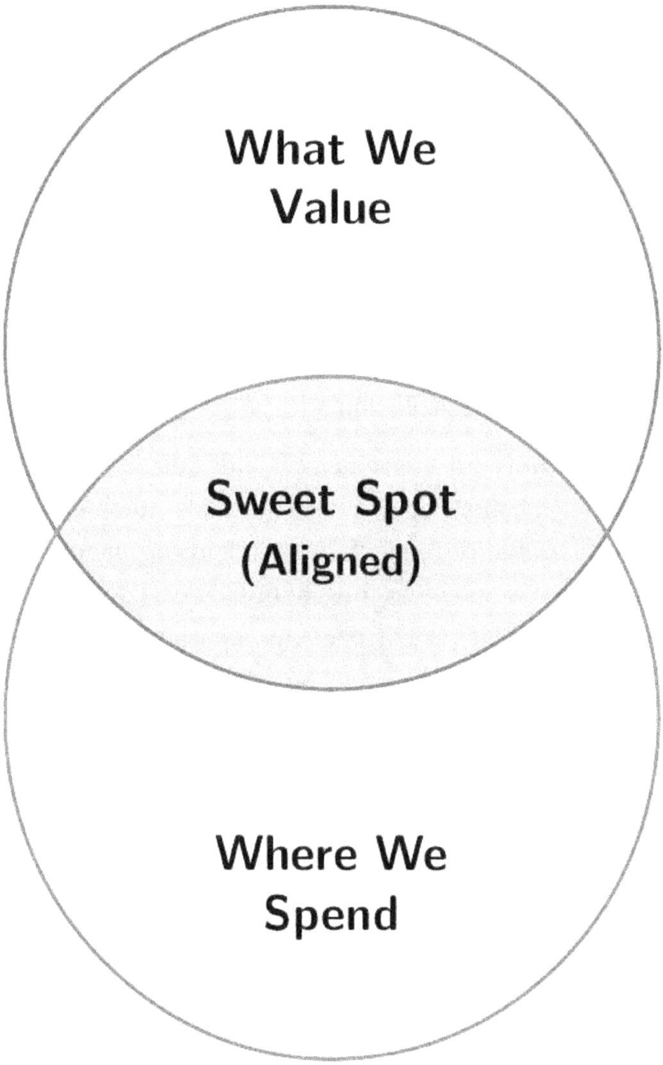
What We Value
Sweet Spot (Aligned)
Where We Spend

A real-world case: the ``yes" that saved a family thousands

Consider Mia and Jordan (composite example, but painfully realistic). Two working parents, two kids (8 and 12), middle-to-upper-middle income. They weren't drowning, but money was always tight in a confusing way: solid salaries, little breathing room, frequent arguments that began as small purchase debates and ended as character judgments ("You're reckless." "You're controlling.").

They assumed their problem was budgeting. What they actually had was a values collision.

Jordan grew up with unpredictable bills and vowed never to feel that fear again. Security was Jordan's north star: emergency fund, stable job, insurance, no surprises. Mia grew up in a house where money was present but joy was postponed. Experiences were Mia's north star: trips, special meals, the sense that life is happening now, not someday.

Without naming these values, every decision became a proxy war. A weekend trip felt like recklessness to Jordan. A "no" felt like deprivation to Mia. The kids learned an unhelpful lesson: money conversations are dangerous; pick a side; ask the softer parent.

They did one short exercise: each adult wrote down the top three money values they believed the family lived by. They didn't match. Then they wrote the top three values they *wanted* the family to live by. Those lists overlapped: security *and* experiences both made the cut, along with learning.

The breakthrough was converting values into rules that weren't personal attacks.

They created three Shared Priorities and paired each with a concrete commitment:

- *Security*: build and maintain a three-month emergency fund; keep a "surprises" line item in the monthly plan.
- *Experiences*: one planned "family memory" per month (cheap counts); one bigger trip per year, funded slowly.
- *Learning*: each kid gets a yearly "skill budget" for lessons, clubs, or gear.

Then they added an equally important rule: *no stealth spending*. Not because spending is bad, but because secrecy turns money into a trust problem. Mia could buy fun things without feeling policed because fun was planned. Jordan could relax because security had a number and a system.

The surprising part: the family didn't just feel better. They spent *less*. Not because they tightened their belts, but because they stopped paying the "conflict tax"—the extra spending that comes from guilt, rebellion, and confusion. When values are clear, "no" gets easier, "yes" gets cleaner, and your money stops leaking through emotional cracks.

Auditing your real values (not the ones you wish you had)

If you want to know your current values, don't start with your philosophy. Start with your transactions. Your bank statement is like a candid photo: sometimes flattering, often honest, always useful.

Pick a recent month and do a quick audit. No spreadsheets required. The goal isn't to punish yourself; it's to reveal patterns.

Here are questions that make the audit interesting instead of grim:

- What category got the most money *and* the most emotion?
- Where do you consistently "treat yourself"—and what feeling are you treating?
- What do you spend on that you never regret?
- What do you spend on that feels like a hangover later?
- What did you pay for because you were tired, late, or overwhelmed?

Then do one brave thing: label two columns.

- *Aligned spending* (it matched a Shared Priority).
- *Unaligned spending* (it didn't match, or it happened on autopilot).

Unaligned spending isn't "bad." It's simply unclaimed. It's money you could redirect without needing a raise or a miracle. Often, families discover that they already have the budget for what they say they want; it's just currently funding stress and friction instead.

Kids can participate in age-appropriate ways. A 6-year-old can circle groceries and "family fun." A 12-year-old can help identify subscriptions no one uses. A teenager can calculate how much takeout costs per month and compare it to a goal (concert tickets, sports fees, a weekend trip). The hidden lesson is more important than the math: adults review decisions.

Adults change plans. Adults don't pretend everything is fine while silently panicking.

Turning values into ``family money rules" kids can actually follow

Values are abstract; kids live in the concrete. "We value security" won't stop an impulse buy at the checkout line. Rules might.

A good family money rule has three features:

- It's short enough to repeat in a store.
- It prevents predictable arguments.
- It points to a value without sounding like a lecture.

Examples (adapt to your life):

- *We wait 24 hours for unplanned buys over $X.* (Impulse control + trade-offs.)
- *We don't buy things to fix a bad mood. We fix the mood first.* (Emotional awareness.)
- *We plan our treats.* (Fun isn't banned; it's intentional.)
- *If it's not in the plan, we write it down.* (Delays without dismissal.)
- *We compare two options before we choose.* (Decision-making muscle.)

One of the most effective rules for families is also the simplest: *We don't decide in the aisle.* The aisle is designed to speed you up and separate you from your future self. If your family values include thoughtfulness, that value deserves a physical boundary: decisions happen at home, with a clear head.

When a kid asks for something, try a values-based response:

> "I like that idea. Let's check if it fits our priorities and our plan."

That sentence does three things at once: it respects the request, it introduces constraints without shame, and it teaches the child that money decisions have a process. Over time, kids start to internalize the process. They'll say, "Is this in my budget?" or "Should I wait?" That's the sound of a future adult who doesn't need a financial rescue mission.

The hard part: values conflict, even in good families

Sometimes the tension isn't between parent and child. It's between two good values.

Security vs. generosity. You want to help relatives or donate to a cause, but you also need to keep your own household stable.

Experiences vs. simplicity. You love trips and activities, but you also crave slow weekends and less chaos.

Convenience vs. health. You're exhausted. The drive-thru solves tonight. It also undermines the food goals you say matter.

When values conflict, the goal isn't to pick one forever. The goal is to make the trade-off visible and temporary: "This month we're choosing convenience because work is intense; next month we'll do a grocery reset." Kids learn that values are not a rigid identity. They're a steering wheel.

It's also worth naming the values you *don't* want running the show. Many families accidentally adopt values like:

- *Avoiding discomfort* (we pay to escape boredom, waiting, or mild frustration).
- *Keeping up* (we spend to avoid feeling behind).
- *Making it up to the kids* (we buy to soothe our own guilt).

Again: no shame. Just clarity. Kids don't need perfect parents. They need parents who can say, calmly, "We got pulled off track. Let's fix it."

A small script that changes everything: ``We're choosing..."

One phrase can rewire your family's money atmosphere: *We're choosing*.

"We're choosing to spend on soccer this season, so we're choosing fewer restaurant meals." "We're choosing to save for a trip, so we're choosing a smaller birthday party." "We're choosing to help Grandma, so we're choosing to pause the home remodel."

This isn't spin. It's truth. Even when money is tight, you are still making choices—and kids feel safer when adults sound like they have a steering wheel, not like they're being dragged behind the car.

If your child asks, "Are we rich?" a values-based answer can be both honest and calming:

> "We have what we need, and we're careful with our choices. In our family, being 'rich' means we can

pay our bills, save for surprises, and still have fun in ways that matter to us."

That answer avoids bragging, avoids panic, and gives your child a definition of wealth that doesn't depend on someone else's sneakers.

Actionable takeaway: write your ``three-line money compass"

Tonight (or sometime this week), do this in ten minutes. Put it on one index card or in a shared note.

1. Write three Shared Priorities: *"In our family, money is mainly for ________, ________, and ________."*
2. Under each one, write one trade-off you're willing to make: *"To protect this, we'll say no to ________."*
3. Write one simple rule you can repeat in a store: *"Our rule is ____________________."*

Then use it once, in real life, the next time a request shows up. Not as a speech. Just as a calm reference point. The index card isn't magic; the repetition is. Over time, your kids won't just learn what your family buys. They'll learn *how your family decides*. That skill outlives every toy dinosaur.

1.2 The Roadmap: What to Teach and When

A five-year-old can remember every single Pokémon evolution but cannot reliably remember where their shoes are. A fourteen-year-old can argue like a junior attorney but "forgets" to bring a lunch they made themselves. This isn't a character flaw. It's brain development: attention, impulse control, planning, and long-term thinking come online in phases. Money skills ride on top of those same brain systems. That's why the most common parenting mistake is either asking too little ("They're just kids") or asking too much ("They should know better by now"). A roadmap keeps you in the sweet spot: ambitious, but fair.

The goal isn't to produce a tiny accountant. It's to grow a young adult who can do four deceptively hard things: wait, choose, plan, and recover. Waiting is the foundation of saving. Choosing is the foundation of budgeting. Planning is the foundation of adulthood. Recovering is the foundation of resilience—because no one gets every money decision right.

A "graduated license" approach works beautifully here. Toddlers don't need keys; they need to sit in the car seat. Later they learn the rules of the road. Eventually they drive with supervision. Money is the same: observation first, collaboration next, autonomy at the end. And crucially, autonomy isn't a birthday gift—it's a skill you build through progressively real practice.

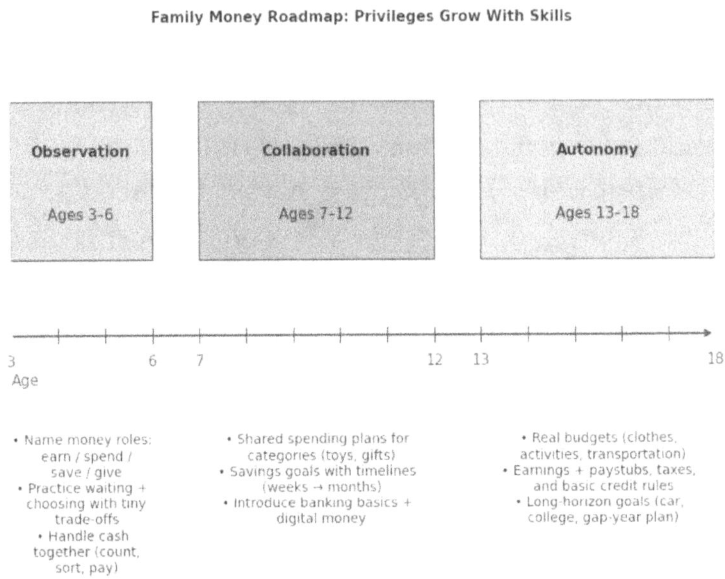

Some parents want a perfect curriculum: lesson plans, quizzes, a neat checklist that guarantees a financially successful adult. Real kids don't cooperate with neat checklists. They mature in spurts. They obsess over weird things. They forget what they know when they're hungry. The better approach is to aim for *milestones* instead of mastery: small, repeatable competencies that stack.

The best money lessons are also the least dramatic. They're woven into ordinary life: paying at a counter, planning a birthday gift, waiting for a sale, choosing between two fun options when you can only pick one. You're not trying to create constant *money time*. You're trying to create a kid who feels competent around money, the same way you want them to feel competent ordering food or emailing a teacher.

Below is a practical timeline you can treat like a menu: pick the parts that fit your child, your values, and your household reality.

Stage 1 (Ages 3--6): Observation

Preschoolers are not ready for "budgeting," but they are ready for the raw ingredients: naming, sorting, choosing, and waiting. At this age, the win isn't a child who can count coins (nice, but optional). The win is a child who can tolerate *not getting everything immediately* without falling apart.

Kids in this stage learn mostly by watching you. If money conversations in your home are tense, whispered, or only happen during arguments, they'll absorb that money equals danger. If money is discussed with calm and clarity—even when you're saying no—they absorb something much more useful: money is a normal tool adults use to make decisions.

What to teach now:

- *What money does*: money is how we trade for things we need and want.
- *The four roles*: earning, spending, saving, giving. Keep it concrete: "We earn money at work. We spend money at the grocery store. We save money for later. We give money to help."
- *Trade-offs*: choosing one thing means choosing not to buy another thing today.

What to practice (tiny and real): Let them hand cash to the cashier. Let them put coins in a jar for a small goal. Let them choose between two snacks at the store. That last one seems trivial; it's actually a masterclass in decision-making under constraint. When you say, "Pick one," you're building the budgeting muscle without using the word.

What privileges make sense: Small amounts of money they can physically see and control. Many families do a small allowance at this age; others don't. Either can work. If you do an allowance, keep it simple and consistent, and don't tie it to perfect behavior. You want the child to practice choices, not to feel like money appears only when they perform.

A surprisingly effective script: "You can't have both. Which one do you choose?" Then—this is the hard part—respect the choice. Don't talk them out of it mid-aisle. If they pick the less sensible option, you didn't fail. You gave them a safe place to learn what regret feels like at low cost.

Stage 2 (Ages 7--12): Collaboration

Elementary and preteen years are the golden zone for building practical money skills. Kids can think ahead a little. They can hold a goal in mind. They can start to understand that "later" is a real place you eventually arrive. They also start to care (a lot) about what peers have. That peer pressure can be a problem, or it can be an advanced course in values: "We don't all spend the same. We spend based on what matters to us."

This is the stage where you move from "I decide, you watch" to "We decide together." Collaboration doesn't mean your child runs the household budget. It means you give them *bounded control*: a category of spending that is truly theirs to manage, with guardrails and coaching.

What to teach now:

- *Spending plans*: a budget is simply a plan for money before it disappears.
- *Saving with a timeline*: "If you save $5 a week, how many weeks until you have $40?"
- *Price awareness*: unit prices at the grocery store, sales tax, and why "two for one" isn't a deal if you didn't need it.
- *Digital money*: money can live on a card or in an app, but it still represents limited choices.

What to practice (real enough to sting a little): Give them a set amount for a category they care about and let them manage it. For a 9-year-old, this might be "book fair money" for the year. For an 11-year-old, it might be a monthly "fun budget" that covers candy, small toys, and impulse wants.

At first, kids will blow it quickly. That's not a sign they're irresponsible. It's a sign the practice is working. Your job isn't to rescue; it's to debrief without shame: "What happened? What would you do differently next month?" The debrief is where the learning lives.

What privileges make sense:

- A savings account you open together (even if the balance is small).
- A simple tracking method they can understand: jars, envelopes, or a kid-friendly app.

- Shared planning for predictable expenses: gifts for friends' birthdays, school trips, or holiday spending.

A concrete example: the "Birthday Season" plan

One family noticed the same money drama every spring: three birthday parties in five weeks, plus class gift exchanges. Their 10-year-old would beg for expensive gifts, and the parents would either overspend to avoid embarrassment or say no in a way that felt harsh.

They created a "Birthday Season" envelope: $60 total. The child could choose how to allocate it: $20 each for three parties, or $10 for one and $25 for another. The rule was simple: the envelope was the envelope.

At the first party, the child spent $25 on a big gift and felt proud. At the second party, the child realized there wasn't enough left for another big gift and panicked. The parents didn't bail them out. They brainstormed: homemade card, a small candy add-on, a thoughtful but cheap item based on the friend's interests. The second gift cost $8 and landed just as well. The child walked away with a skill most adults still struggle with: generosity with limits.

This is collaboration at its best: your kid gets real agency, you keep the stakes manageable, and the family's values show up in a practical form.

Stage 3 (Ages 13--18): Autonomy

Teenagers don't need more lectures about money. They need more *ownership*. Teens are old enough to make meaningful financial decisions—and also old enough to make spectacularly bad ones. That combination is exactly why you want to start autonomy while they still live under your roof.

Autonomy means your teen controls larger categories of spending and experiences the natural consequences. It also means you start teaching the "adult paperwork" that quietly runs life: paystubs, taxes, banking rules, subscriptions, and credit. These topics are often avoided because they feel dry. But to a teen, they can feel like secret knowledge—the behind-the-scenes mechanics of adulthood.

What to teach now:

- *Bigger budgets*: clothing, personal care, entertainment, activities, transportation. Not all at once—add categories gradually.
- *Earnings*: hourly wages, overtime rules, tips, and why gross pay isn't what hits the bank (tax withholding is the portion taken out for taxes).
- *Credit basics*: what credit is (borrowing), what interest is (the cost of borrowing), and the one rule that prevents a lot of pain: "If you can't pay it off in full, you can't afford it."
- *Information literacy*: how to spot scams, predatory offers, and "free trial" traps designed to become paid subscriptions.

What to practice (real life with a safety net): Give your teen a budget for clothing for the semester or the season. If they spend it all in September, they don't get a bailout in October (unless there's a genuine need like a winter coat). They can thrift. They can repair. They can borrow. They can earn. This isn't punishment; it's adult reality in a controlled environment.

You can also practice "bill-like" obligations. Many families choose one: the teen pays for their own streaming service, gas for the car they drive, their portion of the phone plan, or a regular savings transfer toward a goal. The key is consistency: adults aren't financially capable because they know definitions; they're capable because they can handle recurring obligations without drama.

A real-world case: the clothing budget that turned into a life skill

Darius, 15, loved fashion. His parents were supportive but exhausted by the constant shopping requests and the emotional roller coaster: a new trend would drop, Darius would insist he "needed" it, and arguments would flare.

They moved to a seasonal clothing budget: $300 for fall/winter, paid up front into a separate account. The agreement was clear:

- Basics that are truly needs (a warm coat, required school shoes) would be handled separately.
- Everything else—trend items, extra shoes, accessories—came from the $300.
- If he wanted more, he could earn it (chores for extra, babysitting, part-time work).

The first month, Darius spent almost half on one item. He felt amazing for a week, then got anxious when another trend popped up. His parents didn't say "I told you so." They asked, "Do you want to adjust your plan?"

Darius started tracking prices, waiting for sales, and using resale apps. He learned to compare "cost per wear" (how many times you'll realistically use something) without anyone turning it into a lecture. The best part: his confidence grew. He wasn't begging for money anymore; he was managing a portfolio of choices that reflected his values.

Autonomy isn't just about withholding help. It's about giving teens a clear arena where their decisions matter, and coaching them like you would in sports: feedback, strategy, and the occasional hard loss.

What if your child doesn't fit the age band?

Some eight-year-olds are natural planners. Some thirteen-year-olds are still developing impulse control (and some adults never quite get there). Use the stages as a guide, not a diagnosis. The real question is: *How much decision-making can my child handle without being set up to fail?*

A simple test: give a slightly larger responsibility than you think they can handle and watch what happens. If they stumble, shrink the stakes and add structure. If they thrive, expand the privilege. Skill and responsibility should rise together like a well-matched pair of stairs.

Also, life happens. Divorce, job loss, a move, a new baby, health issues—any of these can reset the household's capacity. In stressful seasons, keep the money lessons smaller and more supportive. A child doesn't need you to be a perfect teacher; they need you to be a steady adult.

Two pitfalls that quietly sabotage the roadmap

Pitfall 1: All talk, no reps. If money education is mostly conversations and warnings, kids will treat it like nutrition advice: theoretically true, practically ignored. Skills require repetitions. That means letting kids handle money, make choices, and occasionally feel disappointed.

Pitfall 2: Freedom without a "what happens if." Autonomy without consequences is not autonomy; it's entertainment. If you give a teen a budget but bail them out every time it runs out, you're training them to overspend with confidence. If you give them a debit card but never review statements together, you're training them to ignore the trail money leaves behind.

Consequences don't need to be dramatic. They should be predictable and proportional. "When the fun budget is gone, it's gone." "If you spend your whole clothing budget early, you'll need to thrift or wait." This is how adulthood works, except adulthood is less kind about it.

Actionable takeaway: choose one ``next privilege" and one ``next skill"

Pick one child. Pick one month. Then write two sentences:

- *Next privilege*: "This month, you'll manage ________________ with _____ dollars."
- *Next skill*: "Your job is to practice ______________________ (waiting / choosing / planning / tracking / recovering)."

Set a five-minute check-in each week. Not a lecture—a debrief. Ask: "What did you choose? What surprised you? What would you do differently next week?" If you do nothing else but steadily increase responsibility while keeping the stakes safe, you'll watch something rare happen: your kid will start making money decisions that look like adulthood, not because you forced it, but because they've practiced it enough times that competence feels normal.

1.3 Breaking the Taboo: How to Talk About Money

A kid can smell secrecy the way a dog can smell a dropped snack. You don't have to say the words "money is tight" for them to notice the sharper tone at the grocery store, the way you suddenly go quiet when the credit card comes out, the late-night whisper-fights you think they can't hear. When money is treated like a taboo substance—powerful, volatile, never to be handled in daylight—kids don't become protected. They become imaginative. And a child's imagination, pointed at adult stress, is a factory for worst-case scenarios.

The goal isn't to turn your child into your confidant. It's to make money ordinary enough to discuss, specific enough to learn from, and safe enough that questions don't trigger shame. Kids who can ask about money without fear grow into adults who can ask for help before a mistake becomes a catastrophe. That might be the most underrated financial skill of all.

"Talking about money" isn't one conversation where you reveal your net worth over pancakes. It's a habit of narrating decisions: how you choose, what you trade off, what you care about, and how you recover when you misjudge something. That narration is where values become visible and skills become transferable.

The three traps: secrecy, shame, and oversharing

Most money communication problems fall into one of three traps.

Secrecy is the classic: "We don't talk about money." Parents choose it because it feels polite, or because they grew up with it, or because they worry money talk will burden kids. The trouble is that secrecy doesn't remove the burden; it removes the *explanation*. Kids still sense stress, but now they don't have a story that makes sense. So they write their own. It's rarely flattering.

Shame is stealthier. It shows up as sarcasm ("Must be nice!"), moralizing ("People like us don't waste money"), or humiliation ("You're so spoiled"). Shame doesn't teach budgeting. It teaches hiding. A child who learns that wants are embarrassing becomes a teen who purchases in secret and lies about it.

Oversharing is the trap many thoughtful parents swing into as a correction: "I'll be honest." Honesty is good. But kids don't need the full spreadsheet of adult anxiety. They don't need to know exactly how behind you are, the details of your debt, or the fear you feel at 2 a.m. They need *age-appropriate truth*: enough reality to learn, not so much that they feel responsible for stabilizing the family.

A simple boundary question helps: *Am I teaching, or am I venting?* Teaching builds competence. Venting transfers emotional weight.

Replace ``We can't afford that" with ``We're choosing..."

"We can't afford that" is often technically true, but it has an ugly side effect: it makes affordability feel like an unpredictable verdict handed down by an invisible judge. Kids hear "We might be in trouble," or "Money is scarce," or "Asking is dangerous." Even if your finances are stable, the phrase can create background anxiety.

A better default is: *We're choosing to spend on something else.* It sounds small. It changes everything.

Compare these two responses to the same request:

> "We can't afford that."

versus

> "That's a cool idea. We're choosing to use our money for the beach trip and our savings goal, so we're not buying that today."

The second one does three important things:

- It treats money as a tool for priorities, not a mysterious limitation.
- It models trade-offs without blaming the child for wanting something.
- It reinforces the family's values in real time, which is how values become culture.

Kids don't need you to say yes. They need you to make no sound sane.

Narration: the ``director's commentary" that teaches kids to think

A powerful way to normalize money is to give your child a "director's commentary" track for everyday decisions. You're not asking them to approve the purchase. You're letting them hear how an adult reasons.

Examples, spoken casually:

- "I'm buying the generic cereal because it tastes the same and costs less. That leaves more money for our weekend fun."
- "I'm waiting to buy those shoes. I want them, but I don't need them, and I'd rather not buy things when I'm tired."
- "This is a big purchase, so I'm comparing three options before we decide."
- "I'm setting this bill on autopay so we don't forget. Autopay means the bank pays it automatically each month."

Narration works because it teaches the invisible steps: pausing, comparing, choosing, and committing. Kids don't learn those steps by being told "be responsible." They learn them by hearing what responsibility sounds like.

If you grew up in a house where money talk meant conflict, narration can feel unnatural at first. Start tiny. Narrate one decision per week. Do it in a steady tone. You're building familiarity, not performing a TED Talk.

How to answer the questions kids actually ask

Kids rarely ask, "What is compound interest?" They ask questions that reveal what they're really trying to figure out: safety, status, and belonging.

"Are we rich?"

This question is often a child testing the social map. They're noticing who has what. They're trying to predict whether they'll fit in. They may also be asking, quietly, "Are we safe?"

A useful answer has three parts: reassurance, privacy, and a value-based definition.

> "We're okay. We can pay for what we need, and we make careful choices. Different families spend money differently, and we don't share exact numbers outside our home. In our family, being 'rich' means we can handle surprises, help others sometimes, and still enjoy life."

That answer teaches something better than a dollar amount: wealth as stability and options, not as a scoreboard.

"Why can't we buy it? You have a card."

To a kid, a card looks like a magic wand. This is a perfect moment to teach the difference between *paying method* and *money source*.

Try:

> "The card is just a way to pay. The money still has to come from our bank account. Our bank account is for lots of things: food, the house, saving for later. If we spend it here, we can't spend it somewhere else."

You can make it concrete: "If we bought that $30 toy every time we wanted one, that would be $120 a month. That's the same as _____." Pick a comparison relevant to your life (a family outing, sports fees, a phone bill). Kids understand trade-offs when they're grounded in something real.

"Why do you have to work?"

Underneath this is a child asking, "Why aren't you with me?" You can honor that and still teach economics.

> "I work to earn money, yes. I also work because I'm good at it, and I like contributing. Work pays for our needs and some wants. It also helps us save so surprises don't scare us."

That answer frames work as more than suffering for money. It teaches purpose and planning without pretending every workday is joyful.

"How much do you make?"

There's no universal rule. Some families share numbers; others don't. The more important decision is whether your answer builds healthy boundaries and useful understanding.

If you prefer not to share, avoid making it feel forbidden (forbidden topics become magnets). Try:

> "That's a fair question. We keep exact income numbers private in our family. What I *can* tell you is how we decide what we spend and save, and how you can plan your money."

Then pivot to skills: saving rate, budgeting categories, goal planning. You're not dodging; you're teaching.

A real-world case: the minivan, the neighbors, and the ``not for us" moment

Leah and Sam lived in a neighborhood where cars functioned as social signals. Their two kids (9 and 13) started asking why their family didn't have a newer SUV like "everyone else." The parents could feel the familiar tug: a mix of defensiveness, embarrassment, and a genuine desire to give their kids what's "normal."

They decided to use one specific moment as a family money talk. The old minivan needed repairs. The dealership offered attractive financing on a new model, with a monthly payment that looked manageable *by itself*. That's the trap: most expensive decisions are sold as monthly payments, not total cost.

At dinner, they narrated the decision without turning it into a stress dump.

> "We have two choices. We can repair the van for about $1,200 and keep it, or we can buy a new one and pay about $650 a month for several years. We're choosing the repair. That keeps our monthly bills lower so we can keep saving and still do summer camp."

The 13-year-old pushed back: "But it's embarrassing." The parents didn't shame the feeling. They named it.

> "I get that. It's hard when people have nicer things. But we don't buy things to avoid embarrassment. We buy things because they help us live our priorities."

Then they did something brilliant: they gave the teen a role. "Help us find a reputable mechanic and compare two quotes." Now the teen wasn't just *enduring* a decision; they were participating in adult reasoning. The embarrassment didn't vanish overnight. But it softened into a different emotion: competence. And competence beats status anxiety almost every time.

That's the hidden power of money talk: it turns "my parents won't" into "my family chooses."

How to talk about financial stress without transferring it

If your household is under real strain, you don't need to pretend everything is fine. Kids do better with honesty than with confusing signals. The trick is to tell the truth in a way that maintains safety.

A helpful template:

- Name the reality in simple terms.
- Name the plan.
- Name the boundary: kids are not responsible for adult finances.

For example:

> "Money is tighter than we want right now because __________. We have a plan: we're cutting back on some spending and focusing on essentials. You don't need to solve this. Your job is to be a kid, and our job is to take care of the family."

That last line matters. Many kids, especially sensitive ones, will try to help by shrinking themselves: eating less, asking for nothing, feeling guilty for needing shoes. You can prevent that by explicitly removing responsibility. You can still invite appropriate participation: "We're doing more meals at home. Want to help pick recipes?" That gives them agency without anxiety.

Watch for signs a child is carrying too much: trouble sleeping, stomach aches, persistent worry, sudden perfectionism, reluctance to ask for normal needs. If you see those signs, reduce detail and increase reassurance. The goal is learning, not fear.

The ``ask" rule: how to handle requests without creating a constant debate

Requests are where money talk gets messy. Kids ask at the worst times: when you're tired, in public, surrounded by marketing designed to bypass your brain.

A practical family rule is to separate *asking* from *deciding*. You can welcome the ask without committing to the purchase.

Try:

> "You can always ask. My answer might be yes, no, or not right now."

Then add a process that's predictable. For example:

- Small requests can be decided in the moment.
- Bigger requests go on a list and get reviewed weekly.
- Anything over a certain amount requires a waiting period.

Predictability reduces drama. Kids can tolerate "not now" if they trust there's a real time to revisit it. And you get out of the trap of making decisions in the aisle, where everyone's nervous system is already overloaded.

Language that builds dignity (even when the answer is no)

A no can be delivered in a way that builds maturity or in a way that breeds resentment. The difference is usually dignity.

Phrases that keep dignity intact:

- "That's a want. Wants are allowed. The question is timing."
- "It's not a bad idea. It's just not our priority."
- "Let's price it out and make a plan."
- "We're choosing to save our money for _____."

Phrases that tend to backfire:

- "Because I said so." (Teaches power, not reasoning.)
- "We're broke." (Creates fear; often not even accurate.)
- "You don't need that." (May be true, but it dismisses the desire rather than teaching how to handle desire.)

When you slip (you will), repair quickly: "I snapped. The answer is still no, but I want to say it better." Repair is a money skill too: adults make mistakes and correct course.

Actionable takeaway: adopt one script and one ritual

Pick one script you'll use for the next two weeks:

> "We're choosing to spend our money on __________ right now."

Then pick one low-friction ritual that normalizes money talk without turning your home into a seminar:

- *Car-chat*: once a week, share one money decision you made and why.
- *Receipt review*: after groceries, ask your child to find one good deal and one "oops."

- *Wish-list parking lot*: keep a note where everyone writes down wants; review it on Sunday for five minutes.

Do it consistently enough that money stops being a forbidden topic and starts being what it actually is: a set of choices your family makes on purpose, with calm voices and clear boundaries, even when the answer is no.

1.4 Learning by Doing: The Value of Safe Mistakes

The cheapest financial education you will ever buy is the one that comes with a little disappointment. A toy that breaks. A snack you regret. A game you saved for and then barely play. None of these are tragedies. They are inoculations. They teach the brain, in a way no lecture can, that money is a limited resource and that feelings—excitement, envy, hunger, boredom—are not reliable financial advisers.

Most parents say they want their kids to learn from mistakes. Then the mistake shows up wearing a price tag and tears, and suddenly every protective instinct activates. You want to fix it, replace it, smooth it over, make the disappointment disappear. That urge is love. It's also a hidden trade: you're swapping today's small discomfort for tomorrow's larger one. Because a kid who never practices recovering from a minor money error is a teenager who panics when their debit card runs low, and an adult who reaches for a credit card the moment life gets bumpy.

"Safe mistakes" means calibrated stakes: letting kids feel real consequences while the costs are low and the safety net is still you. The goal isn't to be harsh. The goal is to create a childhood where money has weight, choices have texture, and regret is survivable.

Why experience beats explanation

Adults tend to teach money like it's vocabulary: define *budget*, define *interest*, define *needs versus wants*. Kids nod, repeat, pass the quiz, and then spend their entire allowance on something fluorescent and disappointing.

That's not because they're irrational. It's because the part of the brain that learns from experience is older and stronger than the part that learns from abstract instruction. A child who *feels* the result of an impulse buy is building a mental shortcut: "That didn't end well." This shortcut is what adults call "wisdom," and it's suspiciously hard to download through words.

The paradox is that parents often block the very experiences that would teach wisdom. We rush in with refunds, replacements, extra money "just this once." We do it because we don't want our kids to suffer. But mild, contained suffering is how humans learn to steer. A scraped knee teaches caution on the sidewalk. A broken toy teaches caution in the toy aisle.

The mistake ladder: tiny now, huge later

A 6$mistakeatagesixisabargain.A$60 mistake at age twelve can be useful. A 600$mistakeatagenineteenispainful.A$6,000 mistake at twenty-five can be life-altering. The difference is not morality; it's the size of the arena. Childhood offers a practice field where the numbers are small and the consequences are mostly time and feelings. Adulthood offers a stadium with interest rates, contracts, and long-term obligations.

Interest is the perfect example. Interest is the fee you pay to borrow money (or the money you earn for lending it, like in a savings account). A teenager who has practiced waiting and planning is less likely to borrow impulsively. A young adult who has never practiced waiting will treat interest like a minor inconvenience—right up until it multiplies.

Safe mistakes are how you teach your child that money choices have *echo*. Not endless shameful echoes, just real ones: "If I choose this today, I'm choosing not to have that tomorrow."

A case study: the skateboard that ``didn't count"

Eli was nine and obsessed with skateboards. Not the kind you find at a discount store; he wanted a specific brand he'd seen older kids use. His parents, Nora and Chris, liked the idea of him having a hobby, but they also sensed a familiar pattern: Eli got intensely excited, begged hard, and then moved on quickly once the novelty wore off.

Instead of saying yes or no, they made the decision into a safe-mistake laboratory.

They agreed on a deal: Eli would pay half from his savings jar, they would pay half as a "skill investment." That structure did two things. It gave Eli real skin in the game, and it aligned with a value: learning and persistence matter.

Eli saved for weeks. He watched videos. He talked about nothing else. When the skateboard finally arrived, he used it constantly for about ten days. Then the first big scratch happened.

Then he fell. Then the board started to feel less magical and more like work. Eli drifted back toward video games and asked, carefully, if he could get a different board "that's better for tricks."

This is where the family money culture either becomes real or collapses into convenience. Nora's first instinct was to salvage the joy: maybe they could sell it, maybe they could upgrade, maybe they could find a deal. Chris wanted to shut it down: "Absolutely not, we just bought it."

They chose the third option: a calm debrief and a repair plan.

They sat at the kitchen table with the scratched board between them and asked three questions:

- "What did you expect it to feel like after you had it?"
- "What feels different now?"
- "What do you want to do with the choice you already made?"

Eli admitted the truth: he thought he'd be good fast. He didn't like falling. He liked the *idea* of skateboarding more than the practice.

No one shamed him for that. Instead, they made a plan with real consequences:

- Eli could keep it and commit to practicing twice a week for a month.
- Or he could sell it secondhand and learn what "resale value" means (you don't get full price back).

- Either way, there was no new board until he rebuilt his savings.

Eli chose to practice. He still fell. But something shifted: the board was no longer a fantasy purchase; it was a relationship with an object that required time. By the end of the month he wasn't a skate prodigy, but he was proud. The lesson wasn't "don't buy skateboards." The lesson was "your excitement is not the same thing as your commitment."

That lesson will protect him later from a gym membership he never uses, a financed gadget he stops caring about, or a car payment that looked fun until it became a monthly obligation.

The parent move that makes safe mistakes possible: resisting the bailout

A bailout is when your child makes a money decision and you erase the consequence. Sometimes bailouts are necessary. If your child loses lunch money and can't eat, you don't say, "Natural consequences, sweetheart." If they need shoes that fit, you don't force a lesson about budgeting by letting them go barefoot.

But most bailouts are not about needs. They're about parental discomfort. We hate the crying in Target. We hate feeling like "the bad guy." We hate the sulky silence in the car. We hate the thought that our child might be the only one at the party without the trendy thing.

The solution isn't to become cold. It's to separate empathy from rescue.

Empathy sounds like:

> "Oof. That's disappointing. I get why you're upset."

Rescue sounds like:

> "Fine, I'll buy the other one too."

When you offer empathy without rescue, you communicate two powerful truths: feelings are allowed, and consequences are real. Kids who grow up with that combination become adults who can be upset without immediately spending to escape the feeling.

Regret is data, not a verdict

Many kids treat regret like a character judgment: "I'm bad with money." That belief is dangerous because it encourages giving up. Your job is to reframe regret as information.

A helpful language shift:

- Instead of "That was stupid," try "That was expensive information."
- Instead of "You wasted your money," try "You learned something about what you like *after* you buy it."
- Instead of "See what happens?" try "What will you try next time?"

The aim is not to make them feel better instantly. The aim is to make them feel *capable*. Capable kids don't need you to control everything. They can handle choice.

Designing ``calibrated stakes" by age

You don't need elaborate simulations to create safe mistakes. You just need to match the size of the decision to the child's maturity.

Young kids: Safe mistakes are about immediate trade-offs. If they spend their small allowance on candy, they don't have money for a toy. They learn that "gone" means gone, but the world does not end.

School-age kids: Safe mistakes can involve time. They save for something, buy it, and then realize it wasn't worth it. Time is the sting: "I saved for three weeks for this?" That sting is incredibly educational.

Teens: Safe mistakes should start resembling adult life: subscriptions, budgeting categories, planning for events, earning and saving from a job. The sting might be missing out because they spent early, or having to work extra hours because they committed money to the wrong thing.

The key is to keep the lesson inside a container you can tolerate. If the stakes are so high that you know you'll panic and rescue them, lower the stakes. A smaller arena with consistent consequences teaches more than a huge arena where the rules change when emotions rise.

A ``Reality Fair" at home: the dinner-table simulation that actually works

Some schools and community groups run "Reality Fairs," where kids choose a job, receive a pretend paycheck, and then face realistic expenses: housing, transportation, food, insurance, and surprise events. It's a great concept because it makes adulthood tangible. You can adapt the spirit of it at home without a gymnasium full of booths.

Try a family "Life Night" once or twice a year for teens and preteens:

- Pick a scenario: "You're 22, living in our area, working full-time."
- Give them a monthly take-home pay number (what hits the bank after taxes).
- Give them a menu of expenses with realistic ranges: rent, utilities, groceries, phone, transportation, insurance, entertainment.
- Add a few surprise cards: "Car repair," "Friend's wedding," "Lost your phone," "Hours cut at work."

Then let them make choices and feel the squeeze. Don't turn it into a gotcha. Let it be a puzzle. The learning comes from the gap between what they *assumed* adulthood would cost and what it actually costs.

The best part is the debrief. Ask:

- "What expense surprised you?"
- "What did you cut first, and why?"
- "What did you protect, even when it was tight?"

That last question sneaks in values. Even in a simulation, kids reveal what they care about. That's gold.

The controversy: should you pay kids for good habits?

Some parents pay for reading, exercise, good grades, brushing teeth, chores, you name it. It can work in the short term. Money is a strong motivator. The controversy is what happens long term: if a child learns "I do healthy things only when someone pays me," you may accidentally weaken intrinsic motivation—the internal drive to do something because it matters, not because it pays.

So what's a practical middle ground?

Paying for *contribution* makes sense. Chores are how households function. Paying for some chores (especially extra or unpleasant ones) can be a way to teach earning: work exchanges for money. Even if you don't pay for every chore, you can still connect effort to earning through side jobs: washing the car, helping a neighbor, pet-sitting.

Paying for *personal growth* is trickier. Reading is its own reward, eventually. Exercise protects your health. Brushing teeth prevents pain. If money becomes the reason, you may have to keep increasing the payment to keep the behavior.

If you want to use money as a tool without turning it into a bribe, try paying for *completion of a process* rather than the outcome. Example: "If you finish a four-week practice plan you designed, we'll contribute $X toward your goal." That reinforces planning and persistence, not just obedience.

Or use money as *matching* rather than paying: "For every $10 you save toward the bike, we'll add $5." Matching teaches saving and delayed gratification without attaching cash to every act of self-care.

What to do when the mistake is bigger than you planned

Sometimes a "safe mistake" gets bigger. A teen overdrafts an account. A child buys something in an app. A kid loses cash repeatedly. This is where parents often swing to punishment. But punishment often teaches secrecy, not skill.

A better approach is a repair plan: consequences plus competence-building.

A repair plan has four steps:

1. *Name what happened* without character attacks: "You spent $40 in the game."
2. *Name the impact*: "That money is gone, and it changes what you can do this month."
3. *Create a concrete repair*: "You'll pay it back over four weeks from your allowance/earnings."
4. *Change the system*: remove saved cards, add parental controls, set spending limits, or shift to a prepaid card.

Systems beat scolding. Scolding feels satisfying for five minutes and teaches nothing durable.

Actionable takeaway: run the ``No Rescue, Big Empathy" drill

Pick one small category where your child can make a real choice this week: snacks at the store, a small toy, a book, a game, a birthday gift, a clothing item, an app purchase. Before the choice happens, say one sentence:

> "This is your decision. If you buy it and don't like it later, we won't replace it—but we will talk about what you learned."

Then, when the regret arrives (because it will, eventually), practice the drill:

- *Empathy*: "That stinks. I'd be disappointed too."
- *Reality*: "The money is spent."
- *Repair*: "What's the plan now?"

If you can hold that line with warmth, you give your child a rare gift: the ability to survive a bad choice without panic, shame, or denial. That gift will do more to protect their future than any lecture about interest rates ever could.

1.5 Routines That Stick: Family Money Systems

If you've ever tried to "be good with money" through sheer willpower, you already know the punchline: willpower is a fair-weather friend. It disappears when you're hungry, tired, stressed, or walking past a checkout display that looks like it was designed by a team of behavioral scientists (because it was). The families who seem calm about money aren't morally superior. They have fewer decisions to improvise. They built systems that run even when no one feels heroic.

A family money system is just a set of repeatable routines that turn good intentions into default behavior. Think of it like a household rhythm: the way you do bedtime, the way you handle laundry, the way you get out the door on school mornings. Money can have that same steady beat. Not constant talk, not constant tracking—just predictable moments when money gets handled on purpose, so it doesn't ambush you at 10 p.m. with dread.

The surprising benefit isn't only financial. Systems lower the emotional temperature. When you and your kids know *when* money will be discussed and *how* decisions get made, the random arguments shrink. You're no longer negotiating every purchase like it's a constitutional amendment.

Why systems work (and lectures don't)

Kids don't learn family money norms from your best speech. They learn from repetition: what happens every week, what happens when someone messes up, what happens when the budget is tight, what happens when a request comes at the worst possible time. Systems are powerful because they're emotionally predictable. Emotional predictability is a form of safety, and safety is what allows kids to learn instead of defend.

Systems also help adults. Most overspending isn't caused by ignorance. It's caused by friction and fatigue: forgetting due dates, underestimating irregular expenses, ordering takeout because the fridge looks like a sad science experiment, buying something because you're stressed and want the feeling of control. A system doesn't eliminate emotion. It prevents emotion from being the only driver.

The best family systems are simple, visible, and boring in the best way. Boring means it runs without drama. Boring means it works in February, not just in January.

The "Family Money Loop": earn, decide, review, adjust

Every functioning money system cycles through the same loop:

- Money comes in (earn).
- You direct it (save, spend, give).
- You look back (review).
- You make small changes (adjust).

Most financial chaos happens when the loop breaks. Money comes in, money goes out, and no one looks back until something bounces. Or you review only when you're already upset, which turns "review" into "fight."

Your goal is to make the loop routine enough that it feels like brushing teeth: not thrilling, but oddly comforting once it's a habit.

System 1: The Weekly ``Payday Pause" (10 minutes, maximum)

Pick one day a week when money gets a quick check. It doesn't have to be the actual payday. It just has to be consistent. Ten minutes. Timer on. No spreadsheets required unless you like them.

What happens in the Payday Pause:

- Check the account balance(s) and upcoming bills.
- Transfer money automatically if you can (savings, sinking funds, giving).
- Decide one thing: a small adjustment for the week (fewer drive-thru meals, delay a non-urgent purchase, plan a low-cost outing).

The biggest win here is psychological: you stop being surprised by your own money. Surprises create panic spending ("We're fine, let's just buy it") or panic restriction ("We can't do anything ever"). A weekly pause keeps you in the middle lane: aware, calm, proactive.

Kid version: If your child gets allowance, this is also the moment it becomes a system instead of a random event. Allowance given on a consistent schedule teaches planning. Allowance handed out "when I remember" teaches begging.

A simple family ritual:

> "Payday means save some, spend some, and decide what you're saving for."

Even very young kids can do a two-jar version: "now" and "later."

System 2: The Monthly Family Money Meeting (20--30 minutes, with snacks)

A family money meeting sounds like a beige conference room. It doesn't have to. Think of it as a household check-in where money is one of the tools you use to run life.

Rules that make it work:

- Keep it short and scheduled (same weekend each month).
- Start with something positive (a goal progress update, a win, a moment of good judgment).
- Discuss upcoming events that have costs attached (birthdays, field trips, travel, sports fees).
- End with one decision everyone understands.

This meeting isn't about sharing every adult detail. It's about building competence and trust. Kids who never see adults plan money assume it's either magical or miserable. Seeing planning in action teaches them that money is manageable.

What kids can do by age:

- Ages 5–8: vote on a low-cost family fun option; help choose a saving goal picture for the fridge.
- Ages 9–12: help price out an event ("How much would it cost to bring snacks for the team?"); spot unused subscriptions; brainstorm cheaper alternatives.
- Ages 13–18: manage a category budget (clothing, activities); help compare insurance quotes or phone plans; plan for a big goal with a timeline.

When kids participate, they stop experiencing "no" as arbitrary. They see the machinery behind it.

System 3: Sinking Funds (the secret weapon against ``random" expenses)

Most families aren't wrecked by one huge surprise. They're wrecked by a parade of "small" surprises that were never truly surprises: back-to-school supplies, car repairs, holiday gifts, summer camps, the annual fee you forgot existed until it showed up like a bill-shaped raccoon on your porch.

A sinking fund is money you set aside regularly for an expense you know is coming, even if you don't know the exact date. It's like creating your own personal insurance for predictable life.

Examples:

- Car repairs and maintenance
- Medical co-pays and prescriptions
- Holidays and birthdays
- School fees and activities
- Home repairs

This is where family values become tangible. If experiences matter to you, a "trip fund" makes those experiences possible without debt. If generosity matters, a giving fund prevents last-minute guilt donations that sabotage the rest of your plan.

Kid lesson baked in: Sinking funds teach the concept of "future you." Kids often think adults are just lucky. Sinking funds reveal the truth: adults are planning.

System 4: The ``Open Bill Moment" (quiet transparency)

One reason money becomes taboo is that bills feel like private adult shame. But bills are just the cost of running a life. If you want your kids to become capable, let them see the existence of bills without turning it into anxiety.

Once a month, choose one bill to make visible:

- The electricity bill (and how it changes with season)
- The phone bill (and what a plan includes)
- The internet bill (and how providers price things)
- The car insurance bill (and why it exists)

You don't need to show every line item. You can keep sensitive details private. The goal is to connect "life stuff" to "money stuff" in a calm way.

A good script:

> "This is a bill we pay every month. We plan for it. That's how we keep our life steady."

Kids who grow up around calm bill talk are less likely to avoid their own bills out of fear.

System 5: The ``Ask List" (so requests don't hijack your brain)

Kids ask for things at the exact moment you're least able to decide well: late, hungry, rushed, or in public. A system saves you from making values decisions in the toy aisle.

Create a shared "Ask List"—on a whiteboard, a note app, or the fridge. When a request happens, the default response is:

"Put it on the list. We'll review it on Sunday."

This does not mean you'll buy everything on the list. It means you'll treat requests with respect and process. Kids learn that asking isn't shameful, and you learn that you don't have to decide instantly.

During the weekly review, you can categorize:

- "Yes, now"
- "Yes, later (save for it)"
- "Not for us"

"Not for us" is a deeply underrated category. It teaches kids that values are not universal. Some things are fine for other families and simply not aligned with yours.

A concrete case: the family that stopped fighting by making money boring

Tanya and Marcos had two kids (7 and 14) and a familiar pattern: they were not reckless, but money felt like a constant low-grade argument. It was always the small stuff: Starbucks runs, Amazon orders, "just this once" takeout, the school fundraiser, the birthday gift they forgot until the night before. They didn't have one big spending problem. They had ten tiny ones.

They implemented three routines and agreed not to add more until these stuck:

- A 10-minute Payday Pause on Friday mornings.
- A "Sunday list review" for kid requests and upcoming expenses.

- Two sinking funds: "Kids Activities" and "Car/House Surprises."

The first month didn't feel magical. It felt awkward. They kept forgetting the Friday pause. The teenager rolled their eyes at the Sunday review. But the second month, something happened: the family stopped being shocked by predictable costs. They had money ready for the school trip. They didn't panic when the car needed tires. They didn't get dragged into aisle negotiations because the list existed.

The biggest emotional change came from the sinking funds. When a surprise expense hit, Tanya used to feel angry ("Why does this always happen?"). Marcos used to feel blamed. With a sinking fund, the expense became a transaction, not a betrayal. They still didn't *like* paying for tires. They just didn't treat it as evidence of personal failure.

Money got boring. Their relationship got lighter. The kids learned, without a lecture, that adulthood isn't about never having problems. It's about expecting problems and preparing.

Automation: your future self's best friend

If you can automate something, automate it. Automation is not laziness; it's humility. It's admitting you are a human being with limited attention.

Useful automations:

- Automatic transfers to savings and sinking funds right after payday

- Automatic bill pay for fixed expenses (rent/mortgage, insurance, utilities when possible)
- Account alerts for low balances or large transactions

A helpful rule: automate your *values*. If security matters, automate emergency savings. If generosity matters, automate giving. If experiences matter, automate a small "fun fund" transfer. Then your calendar doesn't have to carry your character.

For teens, automation is also a lesson: you don't build stability by remembering everything. You build stability by building systems that remember for you.

Celebrations that don't turn into spending

Families often celebrate by spending: dinner out, a big purchase, a shopping spree. There's nothing wrong with spending to celebrate. But if celebration always equals consumption, kids learn that happiness is something you buy.

Try celebrating progress in ways that reinforce the system:

- A "goal thermometer" on the fridge that you color in
- A family walk and hot chocolate when you hit a savings milestone
- A low-cost tradition: movie night at home, pancake dinner, board game tournament

Celebration is part of habit formation. It tells the brain, "This behavior matters." It can be joyful without being expensive.

When systems fail (because they will)

You'll miss a meeting. You'll forget the pause. You'll overspend a category. The success metric isn't perfection. It's recovery speed.

A good family repair script:

> "We got off rhythm. That happens. What's one small thing we can do this week to get back on track?"

That sentence teaches an adult skill: when you make a mistake, you don't declare yourself doomed. You re-enter the loop.

Actionable takeaway: build your ``Two-Routine Starter Kit"

Pick only two routines to start this week. More than that looks ambitious and dies by Wednesday.

Choose:

- *One weekly routine*: the 10-minute Payday Pause *or* a Sunday Ask List review.
- *One monthly routine*: a 20-minute family money meeting *or* an Open Bill Moment.

Write them on the calendar like appointments. Add a small reward (coffee, music, dessert) so your brain associates the routine with something pleasant. Then make one decision during the first run: start a single sinking fund with an automatic transfer, even if it's $10.

A family money system doesn't make you immune to stress. It makes you less surprised by life, and less likely to solve every problem with last-minute spending. That calm, repeated rhythm is what kids carry into adulthood: not the memory of what you said, but the muscle memory of what your home did every week.

Chapter 2

The Core Concepts of Money

Think of this chapter as the operating system for your child's financial brain. Before we install the apps-like allowances, savings accounts, or first jobs-we need to make sure the underlying logic is sound. We aren't just teaching math here; we are teaching philosophy. We will move past the simple idea that 'money buys things' to the more mature understanding that money is a finite resource that forces us to reveal what we truly value. Whether your child is five or fifteen, these are the four pillars that hold up the rest of their financial life.

2.1 Money as a Tool: Earning, Spending, Saving, Giving

A lot of kids grow up believing money is something that *appears* right when you need it. The card taps, the receipt prints, the groceries come home. To a child watching from the cart, that looks like magic. And if you grew up with parents who kept money stress private (a loving instinct), you may have accidentally made the magic trick even better.

The antidote isn't a lecture about how hard life is. It's a more useful idea: money is a tool. Tools don't have moods. A hammer isn't "mean" when it hits your thumb; it's just doing what hammers do. Money works the same way. It's a limited resource that flows in, and then you decide where it goes. When kids understand that money is *directed*—not "used up"—they stop being passive consumers and start practicing a skill that will follow them into adulthood: allocation.

Here's the simple reframe that changes everything at the kitchen table: every dollar has four jobs.

- *Spending*: today's life-snacks, games, gas, a birthday gift for a friend.
- *Saving*: tomorrow's you-goals, emergencies, and breathing room.
- *Investing*: future you-money that gets a chance to grow.
- *Giving*: other people-generosity, community, and identity.

You don't need a child to master all four at once. You just need them to see that all four exist. That's how you dismantle the "money equals spending" default setting.

A practical way to make those four jobs visible is the *Four Jars* method. You can use literal jars, envelopes, four labeled boxes, or a banking app with separate "buckets." The physical version is powerful for younger kids because it turns an invisible system into something they can touch. They can hear the clink of coins, see a jar getting heavier, and watch choices take shape.

Now comes the crucial part: the jars are not a moral scorecard. They're a steering wheel. If a child puts too much in Spending one week, you're not trying to catch them being "bad." You're helping them notice the consequence: the Saving jar grows slowly, and the goal takes longer. That's not guilt. That's cause and effect—the cleanest teacher on earth.

Right after you introduce the Four Jars, put the flow on paper (or the wall) so the model sticks even when the jars aren't in the room.

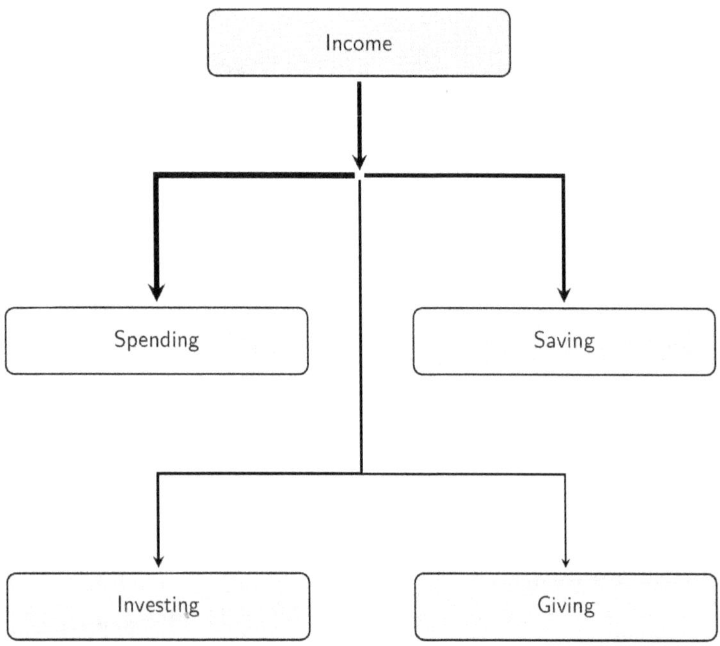

Total Out = Total In

Once kids see the split, they start asking the right questions. Not "Can I have it?" but "Which jar would that come from?" That sentence is the beginning of financial adulthood.

Earning: the antidote to the ATM fallacy

Earning is the inflow. It's the part kids often understand least, because for many families income arrives quietly-direct deposit, autopay, no cash in hand. But earning is where your family's money culture gets built: does money feel like a reward, a right, a weapon, a secret, a scoreboard?

A kid-friendly definition works well: *earning means providing value to someone.* Sometimes that value is work (chores done well), sometimes it's contribution (helping a neighbor), sometimes it's learning (getting paid for a small job because you showed up and tried). The point isn't to turn childhood into a gig economy. The point is to connect money to effort and usefulness, so spending doesn't feel like a disconnected miracle.

One of the most effective conversations is also the simplest: narrate your own earning in plain language. "I get paid because my job helps the company solve problems." "I'm paid hourly, so time matters." "Some months I earn more if I do extra projects." You're not giving them your salary; you're giving them the map: money comes from value, and value takes time.

Chores are where parents get tangled, and it's okay if your family lands differently than your neighbor. Some families pay for chores to teach work-for-pay. Others make chores unpaid to teach shared responsibility ("You live here, you help here"), and then create paid *extra jobs* that are optional. Either approach can work if you're clear about the lesson. The mistake is fuzziness-when kids can't tell whether a task is expected, negotiable, or transactional. That confusion is how you end up with a tiny union negotiator refusing to feed the dog without a contract.

Spending: the jar that talks the loudest

Spending gets all the attention because it produces an immediate feeling: fun, relief, status, belonging. For kids, spending is also a story. The toy isn't just plastic; it's being included. The hoodie isn't just fabric; it's camouflage in middle school.

The spending jar shouldn't be starved into misery, because then it becomes forbidden fruit and kids stop learning-they start scheming. The goal is to make spending *intentional*. Intentional spending means you can enjoy it twice: once when you buy, and again when you remember you chose it on purpose.

A helpful rule for parents: don't treat small spending mistakes like emergencies. If your child blows their spending money on something disappointing, that's not a failure-it's tuition. The adult version of that tuition is far more expensive. A $9 keychain they regret is a bargain compared to a $29,000 car they regret.

What you can do is keep a gentle spotlight on the feeling. "Did that purchase do what you hoped it would do?" "Would you buy it again?" "If you could rewind, what would you pick instead?" Those questions teach a child to evaluate spending not by the price tag, but by the outcome-a surprisingly advanced skill.

Saving: buying time, not stuff

To a kid, saving can sound like a punishment: *don't enjoy life now, maybe enjoy it later*. That framing fails because it's bleak and vague. A better framing is that saving buys *options*. Saving is the jar that turns "I want" into "I can, if I choose."

Start with visible goals, not abstract virtue. A kid saving for a skateboard understands saving. A kid saving for "the future" hears static. The first win is to show that saving is how you make a bigger choice later.

Then introduce the second kind of saving-the kind adults quietly rely on: the "something went wrong" fund. You don't need to scare kids with worst-case scenarios; just let them notice real life. Shoes wear out. A phone charger disappears. A friend's birthday pops up. When you have savings, those moments are inconveniences, not crises.

If you want a sentence that sticks: *savings is what keeps small problems small.*

Investing: planting seeds that can grow

Investing is the trickiest jar because it asks kids to believe in an invisible process. You put money somewhere, and later there may be *more* money. That sounds like a fairy tale-until you make it concrete.

Here's a simple definition that stays true even as products change: *investing means using your money to own something that can grow in value or pay you over time.* With adults, that "something" is often a diversified mix of stocks and bonds through funds. With kids, it can start as a concept: ownership and growth.

A child-friendly way to explain the difference between saving and investing:

- Saving is for money you want *safe and available.*
- Investing is for money you can leave alone so it has time to *try to grow.*

That "try" matters. Investing has risk—meaning the value can go down in the short run. But kids already understand risk in other areas: you practice a new skill and you might fail, but that's how you improve. Investing is the money version of practice.

A concrete family move that works well: choose a small, set amount that goes into "investing" regularly (even $5 or $10). Then track it together monthly, not daily. Daily tracking teaches anxiety. Monthly tracking teaches trends. If your child is old enough, show them a simple line graph and tell them the truth: "Sometimes it dips. We're not in trouble; that's normal. We invest for the long run."

If you'd rather avoid markets at first, you can still teach investing as *building assets*: buying tools to start a lawn-mowing business, supplies to make crafts to sell, or classes that improve a skill. Not every investment is a stock; some investments are in earning power.

Giving: the jar that turns money into meaning

Giving is where money stops being only about the self. Kids are naturally generous in bursts, especially when they can see the person or the impact. What they often lack is a *method*. They'll give impulsively and then feel resentful later ("Now I don't have enough!"), which can accidentally teach them that generosity hurts.

A giving jar solves that. It's pre-decided generosity. It protects a child's kind instincts from the backlash of scarcity. When the giving money is already set aside, the question becomes joyful: "Who do we want to help?" not stressful: "Can we afford to be kind?"

This jar is also where values show up in a way kids can grasp. If your family cares about animals, the giving jar might support a shelter. If your child cares about the environment, maybe they sponsor a cleanup or support a park. If faith is part of your life, giving may include your place of worship. The point isn't the "right" cause; it's that giving is an expression of identity, not a random tax on your happiness.

A real-world case study: Maya's headphones problem

Maya is 12. She gets $15 a week in allowance and occasional extra money for babysitting her younger cousin. One Saturday, she falls in love with $60 wireless headphones. She has $41 in her room-mostly birthday money-and her first instinct is to ask you to cover the difference.

Instead of answering yes or no, you pull out the four jars (or your family's version of them). Maya decides to treat the $15 allowance as her "income stream" and sets a split: $8 spending, $4 saving, $2 investing, $1 giving. It's not perfect, but it's hers.

Now the headphones stop being a drama and turn into a plan.

- She can use her current $41, but that empties her safety cushion.
- If she waits and uses only the saving jar ($4/week), it takes about five weeks to reach $20, and that still doesn't close the gap.
- She has options: take an extra paid job, adjust her split temporarily, or choose different headphones.

Here's where the lesson gets interesting: Maya decides to create a "headphones sprint" for four weeks. She moves her split to $5 spending, $8 saving, $1 investing, $1 giving. She doesn't love cutting spending, but she's choosing it for a reason. She also picks up two extra jobs: organizing the pantry and helping her neighbor weed a flower bed.

By week four, she has the $60. But something unexpected happens—she pauses. The sprint taught her that money is time. Those headphones are not just $60; they're four Saturdays of saying no to other stuff, plus pantry dust and weeds. She still buys them, but she buys them with a different emotion: ownership. Not ownership of the headphones—ownership of the decision.

Two weeks later, the cat knocks the headphones off the counter and they crack. This is the moment that usually creates either entitlement ("You need to replace them") or shame ("I'm so stupid"). With the jars, Maya has a third option: she checks her saving jar, uses part of it for a protective case, and decides to keep the cracked headphones until she can replace them. The jars didn't prevent the problem. They prevented the problem from becoming a family meltdown.

That's the goal: not raising kids who never make mistakes, but raising kids who can recover without panic.

Making the jars work in an actual household

Families abandon systems when the system feels fussy. The Four Jars method wins because it's flexible. You can run it with cash, with a spreadsheet, or with a bank account that uses sub-savings buckets.

A few practical guidelines keep it from turning into a chore:

- *Pick a rhythm.* Weekly works well for younger kids; every payday or twice a month works for teens with jobs.
- *Make the split boring.* Decide a default percentage (or dollar amount) so you're not negotiating every time money arrives.
- *Let kids adjust—with friction.* If they want to move money from Saving to Spending, don't forbid it automatically. Pause and make it a trade: "Okay. What are we trading away?"
- *Keep Investing small at first.* The goal early is the habit and the concept, not the dollar amount.

If you want an easy starting split that doesn't require a calculator, try something like: most into Spending and Saving, a little into Investing, a little into Giving. Then adjust based on age and goals. The most important feature is not the percentages. It's the fact that the split exists before the money gets spent.

One more tip that matters more than it seems: don't use the jars to control your child's personality. If you have a natural saver, celebrate the discipline but also help them practice spending without guilt. If you have a natural spender, don't label them "bad with money." Help them practice waiting in small, winnable ways. Kids tend to become the story you tell about them.

The one-sentence family rule that changes daily life

If your household adopts only one rule from this entire idea, make it this:

When money comes in, we decide its jobs before it goes out.

That's it. Before the vending machine, before the online cart, before the "Can I get this?" moment at the checkout line. Money gets assigned, then spent. It's a small change in sequence, but sequence is where adulthood lives.

Actionable takeaway: a 15-minute ``first paycheck" ritual (even if the paycheck is $5)

The next time your child receives money-allowance, birthday cash, payment for a job-do a short ritual:

1. Put four labels on the table: Spending, Saving, Investing, Giving.
2. Ask your child to choose amounts for each (you can set minimums if you want, but let them feel ownership).

3. Have them physically move the money (or write the amounts into their buckets).
4. Ask one question per jar:

   Spending: "What's something you'd enjoy buying this week?"

   Saving: "What's a goal you'd be proud to reach?"

   Investing: "What's something that could grow if we give it time?"

   Giving: "Who do you want to help, and why?"

You'll walk away with more than a money plan. You'll have a window into what your child cares about-fun, security, independence, generosity-and you'll have given them a repeatable way to turn money from a mystery into a set of choices they can actually run.

2.2 Needs, Wants, and Values

The hardest money fights aren't usually about math. They're about meaning.

A ten-year-old asks for a $60 water bottle. A teen wants the "right" shoes. A seven-year-old melts down at checkout over a toy that will be forgotten by Tuesday. From the adult side of the cart, it's tempting to flatten the whole thing into a tidy lesson: "That's a want, not a need." End of story.

Except it's rarely the end of the story. Kids don't argue because they're confused about the definition of food. They argue because the thing they want is carrying something heavier than the price tag: belonging, comfort, identity, relief, control, fun, status, self-expression. When you treat that entire emotional bundle like a vocabulary quiz-*need or want?*-you miss the real opportunity: teaching your child how to prioritize without feeling deprived, and how to spend without feeling haunted.

The skill you're aiming for isn't "my kid never wants anything." It's "my kid can tell the difference between *I want it* and *I need it*, and can also handle the messy middle without spiraling."

The clean definitions (and why they're not enough)

Start with definitions so the conversation has a shared language:

Needs are the basics required for health and safety: nutritious food, shelter, essential clothing, basic medical care, safe transportation to school, a reasonable ability to learn (which may include some supplies, internet access, or accommodations).

Wants are everything else: nice-to-haves, upgrades, treats, entertainment, convenience, brands, "because it's cute," "because everyone has it," "because it's new."

If your child is young, those definitions alone can be powerful. But as kids grow, the gray area expands fast. A winter coat is a need; a $200 winter coat with a logo the size of a dinner plate is an upgrade. School supplies are a need; premium markers and an aesthetic backpack might be part-need (function) and part-want (style). A phone may be a practical tool for some families; the latest phone is usually not.

The gray area isn't a nuisance. It's the classroom.

Because adulthood is basically one long series of gray areas. Most grown-ups don't go broke buying bread. They go broke buying upgrades: a nicer apartment "for my mental health," a more expensive car "because it's safer," takeout "because I'm exhausted," subscriptions "because it's only $9.99," a closet full of clothes "because I have nothing to wear." That doesn't mean upgrades are evil; it means the decision needs a framework.

The three-category upgrade that actually works

When the need/want label starts a fight, add a third lens: *values*.

Values are the things your family (and your child) cares about deeply, even when nobody is watching. Values are not the same as "what looks good on social media." They're closer to principles and priorities: creativity, kindness, learning, health, independence, adventure, faith, generosity, teamwork, craftsmanship, community, calm.

Values don't magically make money decisions easy. They make them *clear*. They give your child a way to say, "This matters to me," without pretending it's a survival necessity.

Here's the key move: instead of asking only "Need or want?" you start asking two better questions:

- "What problem does this solve?"
- "Which value does this support?"

Those questions turn an argument into a conversation. They also help your child build something most adults never learn: the ability to separate the *story* of a purchase from the purchase itself.

A concrete way to talk about the gray area: function, feelings, and flex

When kids insist a want is a need, they're often mixing three things together:

Function: What does it do? (warmth, transportation, schoolwork, hydration)

Feelings: How does it make me feel? (confident, included, less anxious, excited)

Flex: What does it signal? (status, taste, "I belong here")

All three are real. Pretending feelings and flex don't exist doesn't make them disappear; it just pushes them underground where they control spending in sneakier ways.

So try a calm, curious script:

> "Okay—tell me the function part. What does it do that you need it to do? Now tell me the feeling part. What do you hope it changes for you? And be honest: is there a flex part?"

Kids will sometimes laugh at the word "flex," but they'll answer. And when they answer, you can help them solve the *real* problem with the right tool. If the real issue is feeling included, maybe there are cheaper ways to meet that need. If the real issue is anxiety about being judged, that's not a shopping problem; it's a confidence problem, and money is only playing a supporting role.

Case study: the brand-name sneakers that ``everyone has"

Jordan is 13 and starts middle school where fashion suddenly feels like a currency. He comes home and announces he needs $180 sneakers. Not wants. *Needs*. His parents' first instinct is to shut it down with logic: "You have shoes. Shoes are a want." Jordan hears: "We don't get it. You don't get it. You're on your own."

A more useful approach begins with agreement on the emotion, not the price tag:

> "I believe you when you say this matters. Middle school can be brutal. Let's figure out what's underneath the sneakers."

Jordan admits two things. One: he's worried about being teased. Two: he wants to feel like he belongs with a group that seems confident and put-together.

Now the conversation shifts. Belonging and safety are closer to needs than we like to admit, but the purchase in question is still an upgrade. So his parents put the decision through a values filter:

- Value: *belonging* (real)
- Value: *financial responsibility* (also real)
- Value: *self-respect* (not outsourcing confidence to a logo)

They offer a plan with dignity:

- The family will pay for a solid, comfortable pair of shoes within their normal budget.

- If Jordan wants the premium brand, he can use money from his spending/saving buckets, do extra paid jobs, or ask for the shoes as a birthday/holiday gift.
- They'll also experiment with a "confidence alternative": Jordan chooses one outfit he genuinely likes (not expensive, just *him*), and they practice a simple response to comments ("Yeah, I like these.") so he's not defenseless.

Jordan is disappointed at first. Then something interesting happens: he prices out alternatives, finds a similar style on sale, and decides he can reach the amount by combining gift money with a month of extra jobs. He still wants the sneakers—but now it's a choice with trade-offs, not a demand framed as survival.

Three months later, Jordan tells his mom that two kids at school got teased even *with* the expensive shoes. The shoes weren't a shield. The real shield was having a plan and a spine.

That's what you're teaching: not "never buy the sneakers," but "don't let the sneakers lie to you about what they can fix."

The quiet trap: ``affordable" wants that pile up like snowdrifts

Many families assume wants are the big obvious purchases. But wants usually win through repetition: little things that feel harmless in isolation. A drink after school. A new game skin. A cute pen set. A $7 delivery fee. A dozen subscriptions that started as free trials.

Kids need to learn a counterintuitive truth: a small want, repeated, can bulldoze a big value.

Try a simple household phrase that doesn't moralize:

> "Tiny leaks sink big ships."

Then make it concrete. If your child buys a $4 snack every school day, that's about $20 a week. Over a month, around $80. Over a school year, it can become hundreds. You're not trying to scare them; you're teaching them to zoom out. Kids live in the present by default. Zooming out is a learned skill.

A good follow-up question:

> "If you stopped buying that one thing for a month, what would you be able to do instead?"

That reframes "no" as "trade," which is how adults actually make good decisions.

When a want becomes a need: the role of context

Sometimes parents use "want" to mean "unnecessary," but context changes what's necessary.

A winter coat isn't optional in Minnesota. Reliable internet might be essential for school. A basic phone might be a safety tool for a child who walks home or takes public transportation. A calculator might be required for class. For a child with sensory sensitivities, certain clothing fabrics can move from "preference" to "function" quickly.

Kids notice hypocrisy instantly. If you dismiss something as a want that truly affects their ability to function, they'll stop trusting the categories. The better stance is: the label isn't a weapon; it's a diagnostic.

So keep the conversation open to upgrades that are genuinely supportive:

> "Tell me what happens when you don't have it. What gets harder?"

If the answer is "I don't look cool," that's real but still a want/values conversation. If the answer is "I can't participate in class," "I get blisters," or "I can't sleep," you're in a different category.

How to say no without creating a craving

A hard "no" can be healthy. But the way you deliver it determines whether your child learns self-control or learns obsession.

A "no" that shames ("That's stupid," "Only shallow people care about that," "We're not wasteful like other families") teaches secrecy. Kids who feel judged just start hiding purchases or asking other adults.

A "no" that anchors to priorities teaches adulthood:

> "We're not buying that right now because we're choosing to put our money toward (*value/goal*). You're allowed to want it. Wanting isn't a crime."

Notice what this does. It separates desire from action. Kids need that separation, because desire is inevitable. If they grow up thinking wanting is shameful, they don't become disciplined-they become impulsive in private.

If you want to add a tool, add time. Time turns down the emotional volume:

> "Put it on a 48-hour list. If you still want it, we'll talk about how you'd pay for it."

For younger kids, 48 hours might be too long; try "after lunch" or "tomorrow." The mechanism matters more than the number.

Values as a family language (without turning dinner into a seminar)

Values don't need to be a poster on the wall. They can be a few phrases you repeat so often they become the house accent.

Examples:

- "We pay for what we use."
- "We choose quality in the things we rely on."
- "We don't buy to impress people we don't even like."
- "We take care of our future selves."
- "We help when we can."

The trick is to apply the language to *your own* spending too. If you tell your child brands don't matter, then spend extra for a logo because it makes *you* feel good, they'll see the gap. You don't need to be perfect; you just need to be honest. "I'm paying extra because I really like how it fits and I'll use it for years" is a values-based reason. "Because I want it" is also an acceptable reason sometimes-adults get to have wants-but it lands better when you name it as a want instead of disguising it as necessity.

A practical framework kids can actually use in a store

When you're standing in front of a shelf and your child wants something, there isn't time for a philosophy lecture. Give them a three-question filter they can run quickly:

1. *If we don't buy it, what happens?* (nothing serious / inconvenience / real problem)
2. *Is there a cheaper way to solve the same problem?* (generic brand, borrowing, waiting, repairing)
3. *Which jar does it come from?* (spending / saving / investing / giving)

That last question is the bridge from feelings to reality. It transforms "Do you love me?" energy into "Do you want to allocate your money this way?" energy. It's not cold; it's empowering.

Actionable takeaway: the ``gray-area receipt" conversation

Pick one recent purchase your child cared about-a toy, shoes, a game, a snack run-and do a five-minute debrief. Keep your tone curious, not prosecutorial. Use these prompts:

- "Was it a need, a want, or a values purchase?" (Values purchase means it supports something they care about.)
- "What did you think you were buying *besides* the item?" (fun, calm, belonging, confidence)
- "Did it deliver?" (yes, partly, no)

- "If it didn't, what would you try next time to get the feeling you wanted?" (a different choice, saving for better quality, doing something non-money)

Do this a few times and you'll see the shift: your child starts predicting outcomes instead of just reacting to desires. They'll still want things—because they're human—but they'll be building the rare skill of matching money to real needs, honest wants, and the values they want their life to stand for.

2.3 The Art of the Trade-Off

A price tag is a little white lie.

It pretends the cost of something is only money. But the real cost is always what you *can't* do with that money after it's gone. When a child spends 25 on a toy, they don't just buy the toy-they quietly erase a movie with friends, a bigger goal in the saving jar, a gift for someone else, or a week of not worrying about running out.

Adults call that erased alternative *opportunity cost*. That's a stiff phrase for a very human idea: choosing one thing means rejecting another. Not because you're "bad at budgeting," but because money—and time, and attention—are limited. Scarcity is the rule of life, even in a generous household. The only question is whether your child learns to see scarcity as a punishment or as a steering wheel.

The most useful gift you can give your kid here isn't a formula. It's a habit of mind: whenever you're about to buy something, ask, "What are we trading away?"

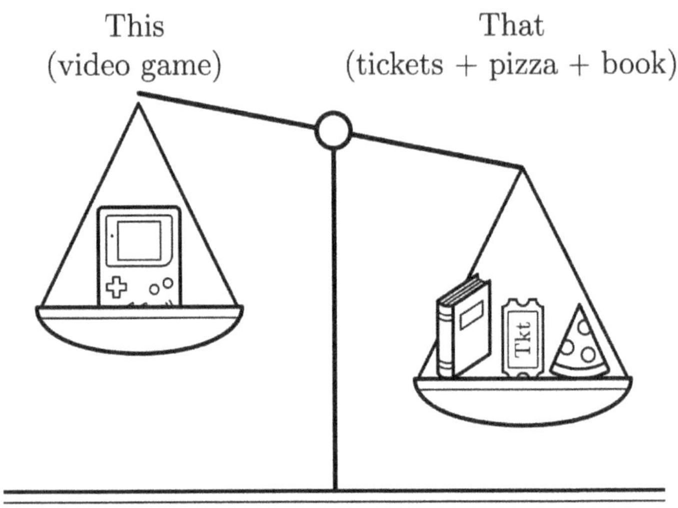

Every purchase is a trade:
This instead of That.

The picture matters because kids don't naturally feel the "that." They feel the shiny "this" in their hand. Opportunity cost is the art of making the invisible alternative visible—without turning every decision into a guilt trip.

Why ``No'' isn't the lesson---*naming the trade* is

A parent saying "no" is common. A parent explaining the trade is rare.

"No, we're not buying that" might stop the immediate purchase, but it doesn't teach decision-making. Kids learn one thing: adults are gatekeepers.

Try something sharper and calmer:

> "We can afford it. We're choosing not to, because we want the money for something else."

That sentence does two quiet miracles. First, it removes shame. Second, it introduces the idea of priorities as an active choice. Even if the "something else" is boring (groceries, rent, the electric bill), you're modeling the truth: money is finite and you're steering it.

Then, if your child is old enough, invite them into the trade:

> "Which would you rather have: this now, or that later?"

You're not asking them to become an accountant. You're teaching them to become a chooser.

The ``This vs. That" habit: a simple script that builds financial instincts

Opportunity cost becomes teachable when you make it concrete. A useful method is to turn any purchase into a comparison:

> "This is ___. That could be ___."

If a teen wants a 120 pair of headphones, "that" might be three months of their streaming subscriptions, or half of a larger savings goal, or a full tank of gas for a month. If a younger child wants a 15 toy, "that" might be two packs of trading cards *later*, a small Lego set, or a friend's birthday present.

Keep "that" in the same emotional neighborhood as "this." Don't compare a toy to "your college education." That's too distant to feel real and turns the conversation into drama. Compare short-range to short-range, medium-range to medium-range.

When kids can feel the alternative, they start doing something powerful: they anticipate regret before it happens.

The secret conversion rate: price tags are made of time

Kids are surrounded by prices, but they rarely connect price to effort. That's how you get a child who genuinely believes a 7 snack is "small" and a 70 pair of shoes is "normal." The numbers float in space.

Time is the anchor.

If your child earns money-through allowance, chores, or a job-translate prices into time and effort. Not as punishment, but as clarity:

> "That toy is three weeks of your saving jar."
> "That game costs two babysitting shifts."
> "That hoodie is six hours at your job."

This does something better than "expensive" ever can. It makes cost personal. Not emotional—personal. Your kid can feel six hours.

For younger kids without paid work, you can translate into *waiting time* or *earning opportunities*. "That's two allowances," or "That's all of your birthday money," or "That's one extra job like washing the car."

Once kids see money as time, they become surprisingly picky. Not because they're stingy, but because they're respecting their own effort. That's dignity, and it's a foundation for later choices about jobs, boundaries, and spending.

Trade-offs aren't just about money (and that's the point)

Opportunity cost isn't only financial. It's a life skill.

If your child spends their whole Saturday gaming, the opportunity cost might be biking with friends, practicing soccer, or just being outside long enough to feel human again. If a teen picks up too many shifts at work, the opportunity cost might be sleep, grades, or mental health. If you say yes to every extracurricular, the opportunity cost might be family dinners that aren't rushed.

You're allowed to say this out loud, because it helps kids see a pattern:

> "Time works like money: when you spend it one place, you can't spend it somewhere else."

That sentence prepares them for adult life better than any lecture about interest rates. It teaches them to notice what their choices are buying—and what they're quietly canceling.

A real-world case study: the console, the camp, and the family decision grid

A family with two kids (ages 11 and 14) faces a classic situation: the older child wants a 400 gaming console. The younger child wants to go to a 350 summer day camp with friends. The parents can't comfortably do both.

This is where many households stumble into a messy outcome: the loudest advocate wins, resentment simmers, and money feels unfair.

Instead, the parents decide to run a simple decision process at the kitchen table. Not a courtroom-more like a planning meeting. They write the problem in one sentence:

> "We have a limited amount to spend on big extras.
> We need to choose what fits our family best."

Then they list alternatives:

- Buy the console this month, skip camp.
- Pay for camp, postpone the console.
- Split the cost: partial family contribution to each, kids cover the rest with their own money or gifts.
- Choose one now, and set a plan for the other later.

Now comes the most important step: criteria. Criteria are the values you're using to judge choices. They can include fun, social connection, learning, longevity (how long the item/experience lasts), health, and fairness between siblings. The parents keep it simple and visible: they pick four criteria and let both kids argue their case *for the criteria*, not just for the thing.

What happens next is the magic of trade-offs: the family discovers that the console is a high-fun, high-longevity purchase that mostly benefits one child, while camp is high-social, high-growth, and time-sensitive. The younger child's friends are going *this* summer; you can't buy "this summer" later.

So they decide: camp gets funded now; the console becomes a shared goal with a timeline. The parents offer to match whatever the older child saves toward the console over the next four months (a simple incentive that respects effort). The older child is disappointed, but not defeated—because the decision feels reasoned and fair. The younger child feels seen, not "spoiled." And the parents didn't have to pretend money is infinite.

That's what a good trade-off conversation does. It doesn't remove emotion; it organizes it.

The trade-off muscle grows in small reps

Parents sometimes wait to teach trade-offs until there's a big purchase, because that's when the stakes feel real. But big-stakes lessons are the hardest place to learn. The emotion is too loud.

Teach the idea in small, low-risk moments:

- At the grocery store: "If we buy the fancy dessert, what are we *not* buying?"
- Before ordering takeout: "If we spend $30 on dinner tonight, what does that trade away this week?"
- During screen-time negotiations: "If you spend your hour on this, what do you give up?"

These aren't meant to produce the "right" answer. They're meant to build the reflex of asking the question. Eventually your child will start asking it without you-sometimes with annoying precision. That's success.

How to handle the classic kid move: ``But I can do both!"

Kids aren't being dishonest when they say they can do both. They're describing a world without limits. Childhood is supposed to be a little like that.

So don't respond with sarcasm. Respond with reality that's still respectful:

> "I wish we could do both too. If we do both, something else gets squeezed. What would you want to squeeze?"

This returns the steering wheel to them. If they say, "We can squeeze savings," you can explore the consequence without shaming:

> "Okay—if savings is smaller, what happens when something unexpected comes up?"

The tone matters. Curiosity teaches. Humiliation just teaches kids to stop talking.

A note for parents: trade-offs reveal values, including yours

Trade-offs can feel uncomfortable because they force you to name what you care about. "We're not doing that because we're paying down debt" can make a child anxious. "We're not doing that because we're prioritizing a family trip" can make one child feel less important.

You don't need to narrate every adult concern. But you can model values without oversharing:

> "We're choosing stability."
> "We're choosing to reduce stress."
> "We're choosing to save for something bigger."
> "We're choosing generosity this month."

Kids learn that money isn't just about consumption; it's about building a life that feels steady and meaningful. That's a much bigger lesson than "we can't afford it," and it's often more accurate.

Actionable takeaway: the two-minute trade-off ritual

The next time your child asks to buy something non-essential, try this quick ritual before you answer yes or no:

1. Ask: "What's the *this*?" (name the item clearly)
2. Ask: "What's one realistic *that*?" (something they actually care about in the same time horizon)
3. Ask: "If you choose this, are you okay with giving up that?"

If they're spending their own money, let them decide-then let the consequence teach, gently. If it's family money, you decide, but name the trade out loud. Either way, you're teaching the skill that keeps adults out of trouble: seeing the second thing on the scale, not just the first.

2.4 Practical Money Math

If you want to watch a child's brain short-circuit, hand them a crumpled $10 bill and point at a price tag that says *$9.99*. They'll do the obvious math-ten is bigger than nine-and feel personally wronged when you explain they still can't buy it.

That tiny moment is bigger than it looks. It's where a kid learns the world has *hidden numbers*. Real-life math isn't the worksheet kind, where everything you need is printed neatly in the problem. Real-life math is detective work. You have to notice what's missing: tax, fees, unit prices, "limited-time" discounts that aren't really discounts, and the sneaky reality that the cheapest item can be the most expensive choice.

The good news: you don't need to turn your kitchen into a classroom. You just need to teach a few practical calculations that show up everywhere—and teach your kid to expect the hidden part.

Sticker price vs. real price: the two numbers that matter

A sticker price is a headline. The real price is the story.

For kids, the simplest way to explain this is: *the real price is what leaves your hand.* If $9.99 turns into $10.79 at the register, then $10.79 is the real price. If a $0 "free trial" turns into $12.99 a month because you forgot to cancel, then $12.99 is the real price. If a $25 concert ticket becomes $41 after fees, that $41 is the real price.

Once your child understands that the number on the shelf is not always the number you pay, they become calmer shoppers.

Not because they're suspicious of everything, but because they stop being surprised.

A useful family phrase:

> "Before we decide, let's find the *real* total."

Sales tax: the invisible plus sign

Sales tax is the first "hidden number" most kids encounter. It's also a perfect training ground because the math is simple and the lesson generalizes: sometimes there's an extra charge that isn't part of the sticker price.

If your child is young, you don't need a lesson on local tax rates. You need an intuition:

> "If it's around ten dollars, tax is about another dollar."

That rule of thumb is close enough in many places to prevent the \$9.99-with-a-\$10 problem, and it builds the habit of leaving wiggle room.

For older kids, you can teach the actual calculation:

$$Tax = Price \times Tax\ rate$$

If the tax rate is 8% and the item is \$50, then tax is $\$50 \times 0.08 = \$4$. Total: \$54.

The lesson isn't "memorize 8%." The lesson is "percent means *per hundred*." Eight percent is eight out of every hundred dollars. That's a concept they'll reuse for tips, discounts, interest, and paychecks.

Tips: the other invisible plus sign (and a values lesson in disguise)

Tipping can be confusing for kids because it looks optional, but it's socially meaningful. This is a chance to teach two things at once: quick percent math and the idea that money choices reflect values.

A fast mental approach:

- 10% is moving the decimal one place left.
- 20% is double the 10%.

So if the bill is $37.50, 10% is $3.75, and 20% is $7.50.

If your child is old enough to notice service work, you can connect it to respect: "Part of eating out is paying the people who made it possible." You're not forcing your child into adult debates about tipping culture. You're simply showing them that math isn't sterile—numbers attach to people.

Percent off: when the store makes math do the selling

Retailers love percent signs because they make our brains lazy. *30% off!* feels like a deal even when the final price is still high, or when you didn't need the item at all.

Teach your child to translate percent off into dollars off. It's empowering because it turns marketing back into math.

Two quick methods:

Method 1: Find the discount, then subtract.

$$\text{Discount} = \text{Original price} \times \text{Percent off}$$

If a jacket is \$80 and it's 25% off: discount is $\$80 \times 0.25 = \$20$. Final price: \$60.

Method 2: Find what you pay. If it's 25% off, you pay 75%. So $\$80 \times 0.75 = \$60$.

Kids like Method 2 once it clicks because it feels like a shortcut.

Then add the lesson that saves real money:

> "A discount doesn't turn a want into a need."

If your child learns to say that sentence internally, you've just inoculated them against a lifetime of "sale" rationalizing.

Unit price: the grocery store's secret language

Unit price is one of the most practical money skills on earth, and almost nobody is formally taught it. It answers a question that sounds simple but is easy to get wrong: *Which one is actually the better deal?*

The trick is that "bigger" isn't always cheaper per unit, and "on sale" isn't always cheaper either. Unit price cuts through the noise by comparing the cost for the same amount (per ounce, per pound, per sheet, per count).

The basic formula:

$$\text{Unit price} = \frac{\text{Price}}{\text{Amount}}$$

If cereal A costs \$4.50 for 18 ounces, unit price is $\$4.50/18 = \$0.25$ per ounce. If cereal B costs \$5.50 for 26 ounces, unit price is $\$5.50/26 \approx \$0.21$ per ounce.

Now you can compare apples to apples (or cereal to cereal).

This does more than save money. It teaches skepticism. Your child learns to distrust packaging and look for the truth in the numbers.

Case study: the ``giant size" trap that fooled a smart kid

A parent takes their nine-year-old, Ellie, to the store to buy snacks for soccer practice. Ellie spots "GIANT SIZE" chips for \$6.99 and a regular bag for \$4.99. She argues hard for giant size: "It's bigger, so it's a better deal."

Instead of correcting her, her parent turns it into a two-minute experiment. They read the ounces:

- Giant size: 15 oz for \$6.99
- Regular: 11 oz for \$4.99

They calculate unit prices together (even with rough math):

- Giant: about $\$7/15 \approx \$0.47$ per oz
- Regular: about $\$5/11 \approx \$0.45$ per oz

Ellie is shocked. The giant size is *not* the better deal. It's close, but it's worse. And then she notices something even more valuable: the "giant size" bag is positioned at eye level with bright colors, while the regular bag sits lower. The store is designing her decision.

That moment is a financial coming-of-age. Ellie didn't just learn division. She learned that smart people can be nudged into dumb choices, and math is how you fight back.

The ``small math" that prevents big arguments: spending plans

Kids often think budgeting means "tracking every penny," which sounds miserable. A better frame is a *spending plan*: deciding ahead of time where money should go so you don't have to renegotiate every decision.

The simplest version is an identity your child can memorize:

$$\text{Money in} - \text{Money out} = \text{Money left}$$

If money left is negative, you have a problem to solve. If it's positive, you can decide what to do with it: save it, invest it, give it, or spend it later.

For teens, this is where you introduce the most adult sentence in personal finance:

> "You don't have to be rich to have a plan; you just have to be honest."

Honesty means including the boring stuff: gas, school lunches, sports fees, haircuts, gifts, subscriptions. Those are the categories that quietly eat a teen's paycheck and then make them feel like money "disappears."

If your teen has a part-time job, help them list predictable expenses and make a default split. It doesn't have to be perfect. It just has to exist before the money vanishes.

Paychecks: the confusing difference between what you earn and what you get

The first time a teen gets a paycheck and it's smaller than expected, they feel cheated. That's not a character flaw—that's a normal reaction to a new kind of hidden number: withholding for taxes (and sometimes other deductions).

A kid-friendly explanation:

> "Your pay has two numbers. *Gross pay* is what you earned. *Net pay* is what you take home after required deductions."

You don't need to teach tax brackets at the kitchen table. You do want them to understand that the advertised wage isn't the same as the cash they can spend. This becomes crucial when they're deciding whether a purchase is "one shift" or "two shifts." They need to think in net pay, not gross fantasies.

A practical habit: if your teen earns hourly wages, help them estimate net pay with a rough rule (for example, "assume 10–20% goes away," depending on circumstances). The point is not precision; it's building the habit of leaving room for the invisible.

Interest: the hidden multiplier that can help you or hurt you

Interest is where money math turns into a superpower—or a trap.

Interest is the cost of borrowing money or the reward for saving/investing it. Same word, opposite direction. Kids find this confusing until you phrase it like this:

> "When you borrow, you pay extra. When you save or invest, you might earn extra."

Two quick examples make it real:

- Borrowing: If you borrow $100 and the interest is 20% for the year, you owe $120.
- Saving: If you save $100 and earn 5% interest, you have $105.

Then add the time factor: interest cares about *how long*. That's the part that makes small amounts powerful over years. You don't need a compound interest lecture; you need a mental picture: a snowball that grows as it rolls. Not a grand metaphor—just an image your child can hold.

If your child has an investing jar, even a tiny one, you can occasionally show them that growth isn't linear. It's lumpy. Some months nothing happens, then later it accelerates. That patience is a learned skill, and math is the language that explains why patience can be worth it.

Fees: the silent budget killer

Fees are the most adult hidden number because they often look small and harmless. Delivery fees. Service fees. Late fees. "Convenience" fees. ATM fees. Account fees. Overdraft fees. Fees are what happen when you pay for speed, forgetfulness, or fine print.

Kids should learn one protective habit:

> "Before you click buy, scroll to the final total."

It sounds almost too simple, but it prevents so many regrets. Make it a family norm when buying online: find the total, including shipping and taxes, before deciding.

For teens, fees are also where you teach a valuable emotional skill: pausing. Most fees are paid in a hurry.

Actionable takeaway: the ``receipt detective" game

The next time you buy something—groceries, a meal, a small online order—hand your child the receipt and play detective for five minutes. Ask them to circle:

- The sticker price(s)
- The hidden numbers (tax, fees, tip)
- The real total

Then ask one question that turns math into judgment:

> "Which hidden number surprised you the most, and what would you do differently next time?"

If they say, "Tax surprised me," they'll start building wiggle room. If they say, "The delivery fee was wild," they'll consider pickup. If they say, "I didn't realize the 'two-for' deal made us buy more," they're learning to resist bait.

That's the point of practical money math: not making your child a human calculator, but making them someone who expects the hidden part, finds it, and chooses with their eyes open.

Chapter 3

Earning, Saving, and Goal Setting

*Money often seems like magic to children-it appears from a wall machine or a card tap and disappears just as quickly. This chapter pulls back the curtain on that magic trick. We will move beyond the passive piggy bank to explore active earning, whether through a structured family allowance, extra chores, or a neighborhood side hustle. More importantly, we'll discuss what happens *after* the money lands in their hand. By turning invisible desires into visible goals and using tangible systems like jars or buckets, we teach kids that saving isn't about deprivation-it's about buying their own freedom, one dollar at a time.*

3.1 Designing the Right Allowance Model

Picture a Tuesday night that goes slightly sideways: homework is half-done, the dog needs a walk, your kid has discovered a sudden, urgent need for *exactly* $12 to buy something their friend has, and you're trying to cook while answering the same question in three different tones: "Can I have money?" Allowance is often treated like a small household detail, but it's really a *systems decision*. The system you pick will quietly shape how your child thinks money *works*: whether money is a right, a reward, a wage, a tool, or a battleground.

There isn't one "correct" model. There *is* a correct model for *your family*: your values, your time, your child's temperament, and what you want money to mean inside your home. If you choose a model on purpose, allowance becomes a low-stakes training ground for skills your child will use forever: planning, trade-offs, self-control, negotiation, reliability, and the ability to recover from small mistakes without shame. If you choose a model by accident, it tends to become either a constant argument or a magic ATM that teaches the wrong lesson very efficiently.

Most allowance systems fall into three philosophies:

No-Strings (a "citizen's dividend"). Your child gets a set amount on a set schedule, simply because they're part of the household. This model separates "being a member of the family" from "being paid." It treats allowance as practice managing money, not a payment for tasks. The household still has expectations, but money isn't the lever used to enforce them.

Earned (a "wage for labor"). Your child gets paid only when they complete tasks. Think of it as a tiny job market: work produces pay. The appeal is obvious: it feels fair, it's easy to explain, and it connects money to effort. The risk is also obvious: children can start treating family contribution like an invoice. ("I'll do it if you pay me.")

Hybrid (a "base + bonuses" approach). This is the practical middle path. Some responsibilities are non-negotiable because they're part of living in a home (baseline contribution). Allowance is provided on a schedule to build budgeting habits, while *optional* paid work exists for extra earning power. Hybrid models try to capture the best of both worlds: steady practice managing money *and* a realistic way to earn more through additional effort.

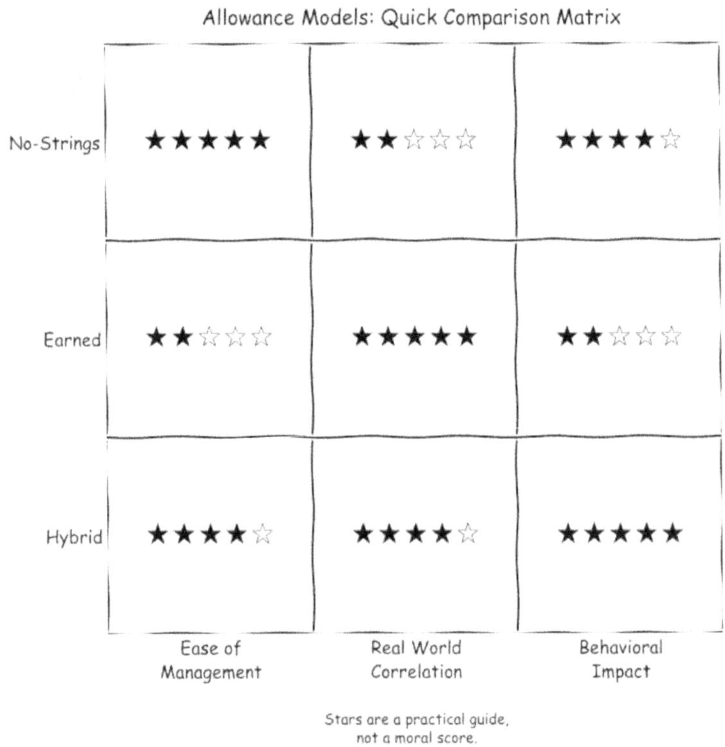

The chart is intentionally opinionated. Not because families need judgment, but because families need *decisions*. The "perfect" model on paper is useless if you won't run it consistently on a normal week when you're tired and your kid is testing the boundaries like it's their job (because, developmentally, it kind of is).

Model 1: No-Strings --- money practice without money drama

The No-Strings model is the cleanest psychologically: allowance is a tool for learning, not a reward, not a threat, not a bribe. It says, "You will have some money to manage, and you'll learn what happens when you manage it well—or poorly—while the stakes are small."

Where it shines is in households that want to keep family relationships out of the marketplace. You don't want your child to see every request as a negotiation. You want them to internalize that some things are done because we live together and care about each other, not because someone paid us.

The biggest misconception is that No-Strings means "no standards." It doesn't. It simply means money is not the enforcement mechanism. If your child doesn't do what's expected, you respond with parenting tools that match the issue: reminders, routines, loss of privileges that are *logically connected* (screen time, playdates, device access), repair work, or extra practice. Not cash manipulation.

The main risk is also the main complaint: some kids interpret unconditional allowance as entitlement. This is why the No-Strings model works best when paired with clear baseline responsibilities and a family culture that treats contribution as normal. Your child can receive allowance because they are part of the family *and* still be expected to participate because they are part of the family. It's only confusing if the expectations are fuzzy or if allowance is constantly renegotiated in response to behavior.

A practical tweak that makes No-Strings feel less like "free money" is to frame it like a *training stipend*. You're not paying them for being alive; you're giving them a small, regular cashflow so they can practice decisions: saving for something bigger, resisting impulse buys, and planning ahead for birthdays and outings. This reframing matters, because kids don't just learn from rules—they learn from the story the rules tell.

Model 2: Earned --- clear cause-and-effect, with a hidden trapdoor

The Earned model has the cleanest math: do work, get paid. Parents like it because it feels motivating and fair. Kids like it because it's legible. If your child is very action-oriented—if they thrive on checklists, visible progress, and immediate feedback—Earned can feel satisfying.

It also maps to a real part of adult life: the labor market. Many adults exchange time and effort for money. There's nothing wrong with teaching that truth early.

The trapdoor is that the Earned model can accidentally turn your home into a tiny corporation where everyone is a contractor. If every contribution is paid, you can end up with a child who won't lift a finger unless the price is right. That's not because your child is "bad"; it's because you trained them on a simple rule: work happens only when it's purchased.

There's another subtle issue: some tasks should not be monetized because they're tied to identity and self-respect. If a child is paid to pick up their own clothes, they might conclude that cleaning up after themselves is optional labor rather than a normal part of being a competent human. It's a small shift in meaning that can become a big habit later.

Earned models also tend to create a management problem for parents. Someone must track the tasks, evaluate the quality, negotiate disputes ("That counts!" "No it doesn't!"), and handle missed payments. It can become a weekly admin job, and the paperwork becomes the curriculum—which is not the curriculum you intended.

If you love the Earned concept, the best way to keep it from swallowing the house is to be very strict about *what qualifies as paid work*. Paid work should be (1) optional, (2) clearly defined, and (3) above-and-beyond normal citizenship in the home.

Model 3: Hybrid --- the adult world, scaled down to kid-size

Hybrid is what many families end up creating anyway, often after running into the limitations of the other two. It recognizes a reality adults live with: some responsibilities are part of the deal (you don't get paid to do your dishes at home), and some work is compensated (your job, side gigs, overtime, special projects).

In a Hybrid system, your child receives a predictable allowance on a schedule so they can practice budgeting and saving. At the same time, there is a menu of *optional paid tasks* or projects that allow them to earn extra when they want to reach a goal faster.

Hybrid tends to reduce conflict because it's emotionally fair in two directions:

- It protects baseline responsibilities from becoming a cash negotiation.
- It gives kids legitimate earning power so they don't feel trapped by "Whatever you decided I get is all I can ever have."

The key is to make the boundary obvious. If the boundary is fuzzy—if you sometimes pay for baseline chores when you're desperate, or you remove allowance unpredictably to punish unrelated behavior—your child learns that the system is arbitrary. Arbitrary systems teach one main lesson: *argue harder*.

A concrete case study: the Martinez family and the $80 headphones

Consider a real-world style scenario (names changed, but the dynamics are common). The Martinez family has a 12-year-old, Lila, who wants $80 headphones. She already gets $10 a week "when everyone remembers." Sometimes it's $10, sometimes it's "we'll get you next week," and sometimes it turns into, "You didn't do your chores, so no money." Everyone feels vaguely irritated.

They decide to design a Hybrid system with two goals: (1) reduce arguments, (2) give Lila a realistic path to earn bigger money for bigger wants.

They create three rules:

1. *Allowance becomes automatic:* $10 every Friday, no negotiation.

2. *Baseline responsibilities are specific:* empty dishwasher daily, pack lunch materials after dinner, and take out trash on Tuesdays. These are not paid.
3. *Optional paid jobs are real jobs:* washing the car ($12), pulling weeds for 30 minutes ($8), organizing the pantry ($15), helping a grandparent with tech setup ($10). Optional means she can say no; paid means it must meet a quality standard.

Now the headphones become a planning problem instead of a drama. Lila can save eight weeks of allowance (slow but simple), or she can accelerate by doing two paid jobs a week for a month. Her parents stop being the "no" machine and start being the "show me your plan" coach.

Two weeks in, Lila does the pantry job and rushes it. Her dad checks and says, calmly, "This isn't finished. Payment happens when the job is done to the standard we agreed." Lila complains. He doesn't lecture. He points to the standard: labels facing out, expired items in a bag, like items grouped. She fixes it. She gets paid. The lesson sticks because it wasn't moralized—it was contractual and clear.

Notice what changed: the family didn't magically remove conflict from parenting. They moved the conflict onto a healthier playing field. Instead of fighting over whether she "deserves" money, they practiced adult skills: clarity, quality, and follow-through.

Choosing a model based on the problem you actually have

Parents often pick allowance models based on what sounds "right." A better approach is to pick based on what consistently goes wrong in your house.

If your home has lots of bargaining and emotional negotiations around chores: No-Strings or Hybrid tends to calm things down because it removes money as the main argument fuel.

If your child struggles to connect effort to outcomes and needs very immediate feedback: Earned (carefully bounded) or Hybrid with an "extra jobs" list can build that cause-and-effect without monetizing everything.

If you struggle to remember allowance and it keeps turning into random cash handouts: any model will fail until you automate the schedule. Consistency is not a virtue-signaling word here; it's the difference between a system and a vibe.

If your child tends to spend everything instantly and then beg for more: any model can work, but No-Strings or Hybrid makes it easier to say, "Your allowance is your practice money. If it's gone, it's gone," without making it feel like punishment.

If your child has anxiety about money or scarcity: predictable allowance on a predictable day is emotionally powerful. It teaches, in the body, not just in the brain, that money can be planned.

The ``micromanagement tax": why simple wins

Allowance systems fail for a boring reason: they require ongoing administration. If the system demands daily tracking, constant judgment calls, and frequent negotiations, it will collapse the first time you have a busy week. And when it collapses, kids learn another lesson: "Rules don't matter; persistence and complaining do."

A simple system has fewer moving parts:

- a set day (and ideally a default method, like cash in an envelope or a transfer),
- a set amount,
- a short list of baseline responsibilities,
- a short menu of optional paid jobs with clear standards.

That's it. You can build sophistication later. Early on, your best friend is a system you can run while tired.

How to price allowance without turning it into a referendum

Parents often ask, "What's the right amount?" Kids often ask, "Why do they get more?" Both questions are really about fairness.

Here's a useful mental model: allowance is not a salary. It's practice cashflow. Practice cashflow should be big enough to allow real choices (so it can teach), and small enough that mistakes don't become catastrophic (so it stays safe).

One approach that keeps you out of endless debate is to tie allowance to what you expect them to cover with it. For example:

- Younger kids might use allowance for small treats, toys, or saving goals.

- Tweens might start covering gifts for friends, extra snacks, or game add-ons.
- Teens might cover a portion of clothing wants, entertainment, or social spending.

Then you can say, truthfully, "We're giving you money because we're giving you responsibility." That's a trade most kids can understand, especially if you start small and increase responsibility gradually.

Quality standards: the secret ingredient that makes ``earned" money educational

Whether you use Earned or Hybrid, quality standards turn paid work into a lesson about real-world compensation: you don't get paid for trying; you get paid for delivering.

Quality standards should be:

- *visible* (a quick checklist or a quick demo),
- *reasonable* (kid-sized, not adult-perfect),
- *stable* (not changing depending on your mood).

This prevents the classic spiral where the child feels cheated ("You're moving the goalposts!") and the parent feels resentful ("I'm paying for sloppy work!"). The standard becomes the referee, not you.

When to dock allowance (and when not to)

One of the fastest ways to make allowance feel chaotic is to use it as punishment for unrelated behavior. If allowance is meant to teach money management, it should be predictable enough to manage.

A cleaner rule is: *don't dock allowance for baseline behavior.* Handle baseline behavior with parenting tools. Handle money with money tools.

The exception is when the money is directly connected to the issue. If your teen agreed to use their allowance for a specific category and then spent it elsewhere, the natural consequence is that you don't rescue them in that category. If they were supposed to buy the movie ticket and blew the money on snacks, they don't go to the movie unless they find another solution. That's not punishment; that's physics.

Actionable takeaway: write a one-paragraph ``Allowance Agreement'' tonight

You don't need a contract with legal language. You need one paragraph that makes the system run without a weekly summit meeting. On a piece of paper (or a shared note), write:

> Allowance is _________ paid every _________.
> It is for practicing spending, saving, and giving.
> Baseline responsibilities that are expected because you live here: _________________________.
> Optional paid jobs you can choose if you want extra money: _________________________.
> Paid jobs are paid only when done to the agreed quality standard.

Then do the most powerful, underrated step: pick the *day* and run the system for four weeks without improvising. Kids don't learn money from speeches. They learn money from the machine you build—and whether it reliably does what it says.

3.2 Chores, Contribution, and Paid Work

A familiar household scene: you ask your child to clear their plate, and they hit you with the line like it's a legal precedent. "How much are you paying me?" They're half-joking. You're half-amused. And then it lands—because if money has been floating around chores for a while, your child is doing perfectly rational math: *If chores equal money, then no money should equal no chores.*

That one sentence is your cue that the family economy needs a boundary. Not a harsh boundary, not a joyless one—just a clear line between two kinds of work that exist in every healthy adult life. There's *citizenship work*: the unglamorous stuff you do because you live somewhere and your existence creates mess, friction, and needs. Then there's *paid work*: tasks someone would reasonably hire out, where the point is results and compensation.

Kids don't naturally understand this distinction because in their world, adults appear to do an endless parade of tasks for no visible pay: dishes, laundry, driving, paperwork, packing lunches, arranging playdates, fixing the Wi-Fi. If you want your child to respect unpaid contribution, you have to make it legible. If you want them to respect paid labor, you have to make it real. The trick is teaching both without turning your home into either a charity or a call center.

The ``family citizenship" idea (and why it's not just about chores)

Start with a simple, sturdy sentence: *"In our family, everyone contributes because everyone belongs."* That is citizenship. It's not punishment. It's not "earning your keep" in a scary way. It's the basic social contract of living with other humans.

Citizenship work is about maintaining the shared space and shared dignity of the household. It includes things like:

- cleaning up what you personally used,
- participating in routines that keep the home functioning,
- small acts that prevent one person (usually the parent) from becoming the default servant.

Notice what's missing: money. That's the point. Money is powerful, and power always changes the meaning of the thing it touches. Once you pay for citizenship work, you're not just handing over a few dollars—you're quietly rewriting the story: "We do this when it's purchased." Some families can run that story and still get decent results, but it often breeds a strange entitlement in both directions: kids feel entitled to payment, parents feel entitled to compliance because they "paid."

If you want your child to grow into an adult who can share a home without constant negotiation—roommates, partners, dorm life, later their own family—citizenship work has to exist as its own category.

The hidden danger of paying for everything: it teaches the wrong kind of motivation

Most parents don't pay for chores because they're greedy capitalists. They do it because they want motivation, and motivation is hard to manufacture at 7:15 p.m. on a school night.

But motivation has *types*. There's internal motivation: "I do this because it matters, it's who I am, it's the standard." And there's external motivation: "I do this because someone gives me something."

External motivation isn't evil. Adults use it every day. You go to work because you get paid. But here's the catch: when you attach money to tasks that are supposed to become part of a child's identity-basic hygiene, basic responsibility, basic respect-you risk teaching them that these things are optional. Not morally optional; *economically optional*. And kids are natural economists. They will immediately ask, "What's the minimum I can do to get the maximum reward?"

Sometimes parents try to fix this by adding more payments, more charts, more bonuses, more punishments. That tends to create the exact problem it's trying to solve: the child becomes even more focused on the reward, and less focused on pride, competence, and belonging.

A cleaner approach is to pay for work that is truly "extra"—work that feels like a job—and keep citizenship work unpaid and expected.

Baseline responsibilities: the tasks that don't get a price tag

Baseline responsibilities are the things a child does not because they're earning money, but because they are learning to be a capable person in a shared home. The baseline should be:

- *small enough to be sustainable,*
- *specific enough to be undeniable,*
- *regular enough to become routine.*

"Sustainable" matters more than ambitious. A baseline list that collapses every week trains kids to ignore it. A baseline list that runs even when you're tired trains kids to trust it.

Examples of baseline responsibilities by age (not as a rigid checklist, but as a menu of ideas):

- Ages 5–7: put clothes in hamper, clear plate, feed a pet with supervision, put toys away before bedtime.
- Ages 8–11: empty dishwasher, make bed, pack school items the night before, take trash to bin, help set the table.
- Ages 12–15: do personal laundry start-to-finish, rotate a bathroom clean, cook one simple meal weekly, handle a routine pet task.
- Ages 16–18: manage a weekly household task without reminders, contribute to meal planning, handle a share of transportation/logistics, run errands with a list and a budget.

The baseline isn't about "earning." It's about competence. You're raising someone who can eventually leave your home and not become a burden to roommates, partners, or themselves.

Optional paid work: where money belongs

Paid work inside a family works best when it feels like the real world in three ways:

It's optional. Your child can say no. That's crucial because paid work is a choice, and choices teach trade-offs. If they choose not to do the paid job, they also choose slower progress toward whatever they want.

It's clearly defined. "Clean the garage" is a trap. It's too big, too vague, and it invites arguments. "Sweep the garage floor and put the bikes in a line" is definable.

It has a quality standard. Not "good job" in the abstract-something observable. "All surfaces cleared, floor visible, trash removed." When a task has a standard, your child learns one of the most important adult money lessons: payment follows *results*, not intention.

Optional paid work is the home version of side gigs. It's the space where kids learn to estimate effort, price their time, deal with boredom, and feel the deep satisfaction of money that was earned cleanly.

Good candidates for paid work include:

- seasonal or occasional jobs (raking leaves, shoveling snow, washing windows),

- deep-clean projects that go beyond daily life (baseboards, organizing a storage closet),
- "parent time-savers" that an adult might pay for (washing the car, cleaning out the fridge—if your kid is old enough to do it safely),
- sibling-help that is genuinely extra (watching a younger sibling while a parent is home but busy, with clear parameters).

The bright line is simple: *paid work should not be the price of basic respect for the home.*

A real-world case study: the Johnsons and the ``Paid to Be Nice" problem

The Johnson family (again, names changed) had a system that looked great on paper: every chore had a dollar value. Make the bed: $1. Put away laundry: $2. Empty dishwasher: $2. Homework done without reminders: $3. "Be kind to your sister": $1.

It worked for about three weeks. Then it got weird.

Their 10-year-old, Marcus, started doing chore math out loud. He would deliberately skip low-paying tasks and only do the ones with the best "rate." He'd offer to "be kind" for money. If he was already being kind, he'd announce, "Do I get paid for that?" His parents found themselves in the absurd position of trying to financially incentivize normal human decency.

The breaking point came when Marcus shoved his sister during an argument and then said, with perfect logic, "Fine, I won't get the kindness dollar today."

That's when the parents realized they had accidentally treated family character as a vending machine: insert dollars, receive behavior.

They rebuilt the system with a boundary:

- Baseline responsibilities became non-paid and non-negotiable: clear plate, clothes in hamper, dishwasher empty on assigned days.
- Allowance became predictable and separate from chore performance (paid weekly).
- Optional paid jobs became a short menu with real standards and real pay (car wash, yard work, organizing a pantry shelf).
- Kindness and school effort moved to their rightful category: expected, coached, and supported—but not bought.

The first week was bumpy. Marcus tested the boundary, because that's what kids do when a system changes. But something surprising happened by week three: he started doing the baseline tasks faster, with less drama. Not because he loved them-he didn't-but because the negotiation was gone. There was nothing to argue about. The baseline was simply the cost of belonging. When he wanted extra money for a game, he chose a paid job. And when he did it, his parents became the "client," not the "judge." The emotional temperature dropped.

That's the quiet win of a good boundary: it doesn't create perfect children; it creates fewer pointless fights.

How to talk about the boundary without shaming anyone

Kids can smell unfairness like smoke. If you introduce baseline responsibilities as "because I said so," you may win compliance but lose buy-in. A better script frames the boundary in adult terms:

> "Some work is family work. It's what we do to take care of the place we live and the people we live with. Nobody gets paid for that—adults don't either. Then there's extra work that goes beyond normal. That's paid, because it's optional and it saves someone time."

If your child asks, "But you get paid at your job," you can say:

> "Exactly. That's paid work. But I don't get paid to pick up my own socks or wash my own dishes at home. Home has its own rules."

That single comparison—job versus home—often clicks. Kids don't need a lecture on economics. They need the categories to feel coherent.

Setting standards without becoming a household inspector

Parents worry that adding standards will make them nitpicky. It can—if standards are vague and emotional. The way out is to make standards *visual and simple.*

Instead of "Clean the bathroom," set a three-part standard:

- Mirror: no toothpaste spots.

- Sink: wiped, faucet shiny.
- Floor: visible, no hair clumps.

Then inspect quickly and neutrally. If it's not done, the phrase is not "This is terrible." It's: *"Not yet. Want me to show you what's missing?"* That's teaching, not criticizing.

For paid work, standards protect both sides. They protect you from paying for half-done work. They protect your child from feeling cheated by arbitrary judgments. A child who believes the system is fair will work harder inside it. A child who believes the system is rigged will either quit or become a lawyer.

What to do when baseline responsibilities aren't happening

This is where many allowance systems get tangled. A child isn't doing baseline responsibilities, and parents reach for the nearest lever: money. But if money is meant to be separate, what's the lever?

Use consequences that are connected to the problem:

- If the kitchen isn't cleaned up, the next fun activity is delayed until it is.
- If school items aren't packed, the child experiences the discomfort of being unprepared (with safety nets for truly important things).
- If laundry isn't done, the child wears what's available.

The goal isn't suffering. It's *cause-and-effect*. Adults don't do dishes because someone hands them $3; they do dishes because the kitchen becomes gross and inconvenient. Natural consequences are the real curriculum.

If your child is consistently failing, it may not be defiance. It may be overwhelm, ADHD/executive function challenges, unclear instructions, or a baseline that's too large. Shrink the baseline, add routine, do side-by-side practice for a week, and then hand it back. Competence comes before independence.

When paid work becomes a ``family business," add adult guardrails

Sometimes kids want to earn money outside the home: babysitting, yard work for neighbors, tutoring, dog walking. That's wonderful—real responsibility with real stakes. It also deserves real guardrails.

First, safety trumps profit. Some jobs are physically risky, some involve strangers, and some require maturity kids haven't built yet. Second, legality exists. Child labor rules vary by age and job type; casual neighborhood work may be treated differently than formal employment, and hazardous work is restricted for minors. Even when something is technically allowed, you still get to decide what's wise.

A practical approach is to make a short "permission checklist" for outside paid work:

- Who is the customer, and how well do we know them?
- What are the hours and location?
- What safety risks exist, and how are they reduced?

- What is the exact promise being made (price, timing, what's included)?
- What happens if the customer isn't satisfied?

You're not crushing entrepreneurship; you're teaching the part that separates real earners from hopeful chaos: clear promises and reliability.

The surprisingly powerful phrase: ``That's not for sale.''

Kids benefit from hearing, occasionally and calmly, that some things are not for sale in your home:

- Basic respect.
- Basic cleanup.
- Basic honesty.
- School effort (not grades, effort).

When you say "That's not for sale," you're teaching values in a way kids can understand: money is a tool, not the ruler of everything. That lesson is protective. It helps kids resist the idea that they can buy their way out of responsibility—or that the world owes them compensation for basic decency.

Actionable takeaway: create a two-column ``Not Paid / Paid'' list and post it

Tonight, grab a sheet of paper and draw a line down the middle.

On the left, title it *Family Citizenship* (*Not Paid*). Write 4–6 baseline responsibilities that are clear, small, and routine.

On the right, title it *Extra Jobs (Paid)*. Write 4–8 optional jobs with simple quality standards and fixed pay.

Then say one sentence that makes the boundary real:

> "Left side is what we do because we live here. Right side is what you can choose to do if you want extra money."

When your child asks for money, you've just given yourself a calm, respectful place to point: not to your wallet, not to your mood, not to an argument—just to a system that makes growing up feel doable.

3.3 First Ventures: Kid Entrepreneurship

The first time a kid sells something to a stranger-cookies, a cup of lemonade, a handmade bracelet-something electric happens. It's not just, "I got money." It's: *"I made money happen."* That's a different identity. Allowance teaches management. A micro-business teaches creation.

And it doesn't have to be a Shark Tank saga. The best "kid businesses" are tiny, local, supervised, and short-lived. They're meant to be a safe lab where a child can practice grown-up skills with kid-sized consequences: noticing what other people want, making a clear promise, delivering on time, counting accurately, and learning that *money in* is not the same as *money you keep*.

A surprising upside: entrepreneurship can calm the constant "can I have..." loop. Not by turning your child into a workhorse, but by giving them a new lever. Instead of pleading for spending money, they can earn it in a way that feels fair to everyone. That shift—*from asking to offering*—is one of the healthiest money transitions a child can make.

What a kid business really sells

Most kid ventures fail for the same reason adult ventures do: they focus on the product, not the customer. Your child thinks they're selling lemonade. The customer thinks they're buying a cold drink on a hot day, a cute interaction, and the feeling of helping a kid who's trying.

That's good news. It means your child doesn't need a groundbreaking recipe. They need to understand the actual job:

- Make it easy for someone to say yes.
- Be pleasant and clear.
- Deliver what you promised.

That's it. Those three behaviors are worth more than perfect signage or a complicated menu.

If you want one simple phrase that makes the whole concept click, try this: *"A business is a promise you keep."* Not a hustle. Not a poster. A promise.

Low-risk venture ideas that work in real neighborhoods

The best first ventures have four traits: low cost, low danger, simple quality standards, and easy supervision. A few classics:

- *Lemonade or cold water stand:* simple, seasonal, and a natural lesson in counting change.

- *Dog walking or pet check-ins (with safety rules):* reliability matters more than enthusiasm.
- *Yard help:* pulling weeds, raking leaves, picking up sticks-clear before-and-after results.
- *"Tech helper":* setting up a phone, organizing photos, teaching basic settings to a grandparent or neighbor.
- *Declutter sale:* selling old toys/books with your help-an early lesson in pricing and letting go.

The goal isn't to maximize earnings. The goal is to maximize learning while keeping risk low.

A case study: Maya's ``Saturday Dog Loop"

Maya is 13, wants $120 for a used bike she found online, and she's tired of waiting. She likes dogs, her neighborhood is full of them, and she convinces her parent to help her try dog walking.

They make it small and safe. Maya doesn't walk unfamiliar dogs at first. She starts with two families they already know from school. She offers a simple package: Saturday mornings, 30-minute walk, text a photo mid-walk, $15 per dog.

The first week is chaos in miniature. One dog refuses to move. Maya is late to the second house. She forgets poop bags and has to run back. She gets home sweaty and annoyed and says, "This is harder than it looks."

Which is exactly the point.

Her parent doesn't rescue her with a lecture. They treat it like a business debrief:

- What went wrong that you can control next time?
- What will you do differently?
- What will you tell your customers?

Maya creates a checklist: bags, water, spare leash, phone charged. She builds in a 10-minute buffer. She texts, "I'm running 5 minutes late—on my way." The second week goes smoothly. By week four, she has three dogs and a waiting list for Saturdays.

Then the real lesson arrives: Maya proudly announces she "made" $180 in a month. Her parent asks one question: "How much did you *keep*?"

Maya looks blank. She hadn't counted treats she bought to make dogs like her, a new leash she insisted she "needed," and the reflective vest. It wasn't a failure. It was an opening. Now she's learning the adult version of money: *profit*.

Revenue vs. profit: the lesson hiding under every lemonade stand

Kids (and plenty of adults) naturally focus on the satisfying number: the cash collected. That number is called *revenue*—money coming in. Revenue feels like success because it's visible and immediate.

But revenue isn't the win. *Profit* is the win. Profit is what's left after you pay the costs required to make the sale. Costs can be obvious (lemons, cups) or sneaky (markers for the sign, dog treats, transportation, replacement supplies).

The most empowering thing you can teach early is this simple equation:

Money In (Revenue) – Costs = Money to Keep (Profit)

And then you can add an even more grown-up layer without making it heavy: profit is what lets a business grow, save, or weather a mistake. Profit is breathing room.

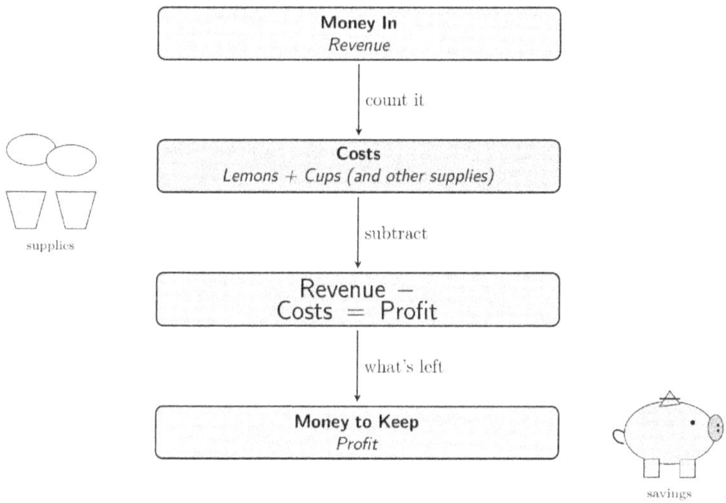

Once your child understands the profit equation, they start asking better questions. Not "How much did I make?" but "How much did it cost me to make that?" That's the beginning of financial maturity.

Pricing without overthinking: the ``easy yes" rule

Pricing is emotional. Price too high and you feel awkward. Price too low and you feel taken advantage of. Kids feel this intensely because they're new to the idea that their time has a dollar value.

A simple first rule is the "easy yes" test: set a price that makes a reasonable customer think, *Sure, why not?* The goal of a first venture isn't to extract maximum dollars; it's to get reps—repetition, practice, feedback.

Then teach the most important pricing insight adults learn the hard way: *price is allowed to change*. If a child sells out instantly every time, the price might be low. If no one buys, it might be high—or the offer isn't clear—or the timing is wrong. Adjusting price isn't "being greedy." It's learning from reality.

A practical way to make pricing less personal is to frame it as an experiment:

> "Let's try $1 a cup for the first hour. If it sells fast, we can test $1.50. We're not changing who you are; we're changing the data."

Kids relax when pricing becomes a test instead of a judgment of their worth.

Customer promises: the skill that matters more than hustle

Adults love to praise "hustle." Kids hear that and think effort is the main ingredient. Effort matters, but customers don't buy effort. They buy outcomes: the dog gets walked, the lawn gets raked, the drink is cold, the tutoring session actually helps.

So put your child's attention on promises:

- What exactly are you doing?

- When will it be done?
- What will it cost?
- What will you do if something goes wrong?

You can turn this into a tiny script they practice before knocking on a neighbor's door:

> "Hi, I'm Maya. I'm offering 30-minute dog walks on Saturdays. It's $15, and I'll text you a photo during the walk. Would you like a spot this Saturday?"

This is gold. It teaches confidence without pushiness, clarity without begging.

Reliability is also where entrepreneurship becomes character-building in a way allowance never can. When your kid has promised to show up, they feel the weight of being counted on. That's the muscle that later makes them a reliable employee, friend, partner, and parent.

Recordkeeping: the boring habit that makes money feel real

If you want your child to take their venture seriously, don't start with a complex spreadsheet. Start with a pocket notebook or a single page on the fridge. The record should capture:

- date,
- what was sold/done,
- money in,
- costs,
- money kept.

This is where the profit equation becomes lived experience. It also quietly teaches honesty and accountability. A child who learns to track money early is far less likely to drift into "I have no idea where it went" later in life.

If your child's business becomes surprisingly successful, recordkeeping also protects them. In the U.S., self-employment income can create tax filing obligations once net earnings reach certain thresholds (a commonly cited federal threshold is $400 in net self-employment earnings). You don't need to scare kids with taxes, but you can normalize the adult idea that money has rules—and keeping records is how you stay calm inside those rules.

Safety and boundaries: the parent's job in the ``kid CEO" era

When kids hear "business," they can get intense fast. They'll want to expand, scale, and recruit their siblings like they're building an empire. Your role is to keep it healthy.

Some non-negotiables:

- No entering strangers' homes without a parent-approved plan.
- No handling hazardous tools or chemicals.
- No jobs near traffic without supervision.
- Clear rules about which neighbors are "yes" customers.

If your teen wants formal employment (a job at a business), child labor laws and safety rules may apply depending on age and role. Even when the work is legal, the most important question is still parental: *Is this safe, and is this environment good for my kid?*

Entrepreneurship should expand a child's world, not expose them to adult risks too early.

The most valuable failure: when the business doesn't work

A kid who tries a venture and fails is not behind. They're ahead. They have already done something many adults avoid: they tested an idea in public.

If your child runs a lemonade stand and makes $6 after buying supplies, resist the urge to "top them up" so they don't feel disappointed. Disappointment is part of learning, as long as it's not humiliating. Help them do a warm, practical post-mortem:

- Were we in the right spot?
- Was the offer clear?
- Did we price it in a way that made an easy yes?
- Were the costs too high?

Then—crucially—help them try again quickly, with one change. Iteration teaches resilience. Rescue teaches fragility.

Actionable takeaway: run a ``one-hour business sprint" this weekend

Pick a venture your child can do in one hour with you nearby (lemonade, cold water, a mini yard-help offer to one neighbor). Before you start, write four lines on paper:

- *Promise:* What exactly are we delivering?

- *Price:* What will we charge?
- *Costs:* What will we spend first?
- *Profit:* After we subtract costs, what do we keep?

Then do the sprint. Count the money together, subtract the costs together, and let your child hold the profit in their hand for a beat longer than usual. That pause is where the lesson sinks in: not "money appears," but *value creates money*—and keeping promises is how you earn it again.

3.4 Visualizing Savings: Jars and Buckets

Kids can understand the idea of saving in the same way they can understand the idea of "later." They understand it *in theory*, like they understand that vegetables are good for you. But the future is foggy, and money in a wallet (or a number in an app) is dangerously present. A ten-dollar bill doesn't whisper, "Someday." It shouts, "Candy. Now."

That's why the most effective savings tool for kids is almost embarrassingly low-tech: separation you can see. When you give money a physical home—jars, envelopes, a labeled pouch, a dedicated bank "bucket"—you turn saving from an abstract virtue into a concrete action. The brain relaxes. The decision becomes simpler. A child doesn't have to constantly *remember* their goal. The container remembers for them.

Adults do this too, even when we pretend we don't. We call it "budgeting," "accounts," "sinking funds," "envelopes," "mental accounting." Kids call it "my jar." Same idea, different packaging: money behaves differently when it has a name.

Why separation works (and why willpower usually doesn't)

If you've ever watched a child hold cash, you've seen how immediate it feels. There's a psychological reason: cash is tangible, and tangibility makes rewards feel bigger and closer. A savings goal, on the other hand, is invisible. It lives in words.

Separation flips that. A "Save" jar makes the future visible. A "Spend" jar makes the present limited. And a "Give" jar introduces the radical idea that money can be used to change someone else's day. You're not just teaching restraint—you're teaching categories, priorities, and choice.

This is also why "Just don't spend it" is a weak instruction. It's like telling a kid to keep a balloon in the air without letting it touch the ground. They can do it for a while, but it's exhausting, and one moment of distraction ends the whole thing. Containers remove some of the load from willpower and hand it to the environment. In parenting, that's often the difference between a lesson that sticks and a lesson that becomes a lecture.

The Three Jar System: Spend, Save, Give

For younger kids, the Three Jar System is the sweet spot: simple enough to run, rich enough to teach. The labels matter. They are not "now," "later," and "charity" (too abstract). They are action words:

- *Spend:* money for small choices now.
- *Save:* money for something bigger later.
- *Give:* money you use to help, share, or contribute.

You can start as early as kindergarten, not because your five-year-old will become a miniature investor, but because the habit you're building is *allocation*. Allocation is the grown-up skill of deciding where money goes *before* it disappears.

Here's the key move: don't make the jars theoretical. Make them operational. When money comes in-allowance, birthday cash, tooth-fairy money-it gets split immediately. Not "sometime." Not "after we buy this one thing." Immediately. The moment becomes a tiny ritual: count, split, drop in, listen to the clink. The clink is the curriculum.

Choosing a split that teaches without punishing

Parents sometimes get stuck here because it feels like setting a moral policy: "How much should my kid give?" "How much should they save?" The best answer is the one you can keep doing without turning it into a weekly argument.

A practical starter split is 50/40/10:

- 50% Spend (because kids need the fun of agency),
- 40% Save (because goals need momentum),
- 10% Give (because generosity is a skill, not a personality trait).

Another option is 60/30/10, or 70/20/10 for kids who need more immediate spending power to stay engaged. The point isn't the exact percentages; the point is that saving and giving become normal. You want your child to grow up thinking, "Of course some of this gets saved. Of course some of this gets shared." That mindset is more valuable than squeezing a few extra dollars into the Save jar and triggering rebellion.

One note that's surprisingly important: let the "Give" jar be directed by the child whenever possible. If giving is always your preferred cause, it becomes your spending, not theirs. If your child wants to buy dog food for the shelter because they love animals, that's a win. If they want to contribute to a school fundraiser because their friend is doing it, also a win. The money is teaching empathy through choice.

A case study: the Nguyen family and the vanishing birthday money

The Nguyen family had the classic problem: their 8-year-old, Ethan, would receive birthday money from relatives and it would evaporate in a weekend. Not because he was careless-because he was eight and the world is a candy store.

His parents tried talking about saving. Ethan would nod seriously, as if he were receiving a sacred adult secret. Then he'd walk into a store, see a bright box, and the secret would be gone.

They switched to jars.

On Ethan's birthday, he received $60. They sat at the kitchen table and said, "We're going to do the jar split first, then you can decide what to do with the Spend jar." They used 50/40/10:

- Spend: $30
- Save: $24
- Give: $6

Ethan's body language changed when the money hit the jars. The Save jar was suddenly a growing object, not an idea. He could see it. He could shake it. And because he still had $30 in Spend, he didn't feel trapped. He chose a small toy and some trading cards and still had money left-an experience of agency without the post-purchase panic.

Two weeks later, Ethan wanted a $45 LEGO set. Before, this would have triggered begging or bargaining. This time, he walked to the kitchen and stared at the jars. His Save jar had $24. He did the kid version of budgeting: "If I wait until next time, I can get it." Then he did the truly magical part: he *waited*. Not perfectly, not without complaining—but he waited, because the plan was visible.

The jars didn't make Ethan more virtuous. They made the future easier to hold in his mind.

Troubleshooting the most common jar problems

"My child keeps raiding the Save jar." This is not a reason to abandon jars; it's a reason to add a rule that mirrors reality: money has friction when you move it. Put the Save jar somewhere slightly inconvenient (a high shelf, a closet), or use a jar with a simple twist lid that requires permission. You're not being controlling; you're adding the same barrier adults use when they put savings in a different account.

"My child only wants to Spend." That's normal, especially early. Keep the Save split small enough that it doesn't feel like confiscation, then tie saving to a concrete goal. Saving without a goal feels like dieting without a reason. Saving *for* something gives the brain a picture to chase.

"Relatives give money and tell my kid to spend it." Smile and accept the gift, then quietly apply your household system. You're not disrespecting relatives; you're respecting your child's education. If it helps, you can say to your kid later, "They meant, 'Have fun with it.' Fun can include saving for something big."

When kids get older: buckets, not jars

As kids move into middle school and high school, they outgrow the cute jar aesthetic. They want autonomy and privacy. They may also start earning money in bigger chunks from babysitting, lawn work, or a small venture. At that point, the same principle—separation—needs an upgrade.

Enter "buckets."

A bucket is simply a labeled category where money lives. It can be:

- a dedicated envelope in a drawer,
- a separate section in a binder,
- multiple accounts (checking + savings),
- or "sub-savings" categories inside one account tracked on paper or in a notes app.

The teenage brain is capable of more complexity, but it is also more vulnerable to impulse spending because temptations are everywhere and social pressure is real. Buckets are a way to protect future goals from present cravings without requiring constant parental policing.

A teen-friendly bucket setup often looks like:

- *Spending (weekly use):* snacks, outings, small wants.
- *Goal Fund (specific):* headphones, prom outfit, sports gear, a trip.
- *Buffer (small emergency stash):* forgotten lunch, replacing a lost item.

Notice what's missing: a dozen categories. Complexity is not maturity. Clarity is maturity.

The "accidental spending" problem (and why buckets solve it)

Many teens don't "blow their savings" in one dramatic moment. They leak it out through tiny purchases that feel harmless: a drink here, an in-app purchase there, an impulse snack, a late-night convenience store run. Each decision feels small; the total is huge. This is how adults derail savings too.

Buckets solve this by giving money a job before it hits the teen's spending stream. When a teen says, "I'm saving for a car," but all their money sits in one account, the car fund is constantly exposed to a thousand small temptations. If money for the car is moved into a separate bucket—separate account, separate envelope, or even a strict tracking category—the car becomes harder to steal from.

Harder is the point. A good system doesn't require hero-level self-control. It requires a few practical obstacles between "I want" and "I bought."

Making bucket transfers feel automatic (the habit that changes everything)

The most powerful moment to build a savings habit is the moment money arrives. For teens with a bank account, you can mimic the jar ritual using a simple rule: *when money comes in, buckets get filled first.*

If your teen earns $80 babysitting, the script is:

> "Before you spend anything, decide what goes to Spending, what goes to your Goal Fund, and what goes to Buffer."

They can choose the percentages within a structure you agree on. For example:

- Minimum 20% to Goal Fund,
- Minimum 10% to Buffer until it reaches a set cap (say $40),
- The rest to Spending.

The numbers matter less than the sequence. Sequence teaches identity: "I'm the kind of person who pays future-me first."

A teen case: Sofia, the concert ticket, and the bucket that prevented regret

Sofia is 15 and obsessed with a band that's coming to town. Tickets are $95, and her friends are going. Sofia has money in her checking account from occasional babysitting-about $160. She also has a goal: save $300 by summer for a short trip with her cousin.

In a single-account world, Sofia buys the ticket, then a shirt, then snacks at the mall with friends, and later she feels the quiet dread: "I can't do the trip." The dread turns into a request for help, which turns into a family argument, which turns into, "Why can't you manage your money?"

In a bucket world, Sofia has a Goal Fund bucket that already holds $120 of the $160. Only $40 is in Spending. When she goes to buy the ticket, she can't do it without consciously transferring money out of the Goal Fund—and that pause is everything. She thinks for a day. She decides to do one extra babysitting job and sell a couple of old items online. She buys the ticket a week later without sabotaging the trip goal.

Buckets didn't deprive her. They prevented regret by making trade-offs visible before the purchase, not after.

How to keep jars and buckets from becoming control tools

The danger with any system is that parents start using it to force outcomes: "You *must* save 80%," "You can't spend any of that," "Giving has to go to what I approve." When that happens, the child learns compliance, not competence. The goal is to raise someone who can manage money when you're not there.

So keep two things true:

- The child has real choice within the structure.
- The structure is stable enough that choice has consequences.

If you change the rules every time your child makes a decision you don't like, you're teaching them that the system is fake. A fake system produces fake learning. A real system produces real skill.

Actionable takeaway: set up a "money landing pad" and do a 3-minute weekly reset

Pick the version that fits your child's age:

For younger kids: three jars labeled Spend/Save/Give. Choose a split you can live with. Do the split immediately whenever money arrives.

For teens: create three buckets (Spending/Goal/Buffer) using envelopes, a tracking note, or separate accounts. The exact tool matters less than the labels and the separation.

Then add one routine that makes the whole thing stick: a weekly 3-minute reset. Same day each week, count what's in each jar or bucket, and ask one question:

> "Does this still match what you care about right now?"

That question teaches the real adult skill under all of this: money is not just math. Money is a map of priorities—and a labeled container is how kids learn to read it.

3.5 Turning Dreams into Plans: Goal Setting

There's a particular kind of kid sentence that can melt a parent's resolve in three seconds: "I've wanted it *forever*." Forever usually means ten days. But the feeling is real. Wanting is loud. Waiting is quiet. And money—especially when it shows up in small, irregular bursts—can make the loud part win almost every time.

Goal setting is how you teach a child to turn the volume down on "I want" and turn the volume up on "I'm getting." Not by killing desire. Desire is useful; it's fuel. The skill is learning to build a track for the fuel to run on, so it doesn't just flare up and vanish in smoke.

A good money goal is more than a wish with a price tag. It has a picture, a number, a timeline, and a plan you can actually follow on a normal week. When kids learn to do that, they're not just learning patience. They're learning something deeper and more adult: *the future can be negotiated with.*

Start with the picture, not the math

Adults love to begin with numbers because numbers feel responsible. Kids begin with feeling. If you skip the feeling, the math won't stick.

So begin with a picture. Literally.

If your child is younger, have them draw the thing they want. If they're older, have them save a photo in a notes app or print a picture and tape it somewhere. The point is not artistry; the point is *emotional clarity*. A goal that lives only as words ("a new game," "some shoes," "a skateboard") has weak gravity. A goal you can see has pull.

Then ask one question that quietly upgrades the entire conversation:

"What will this let you *do*?"

A bike means freedom. A game might mean belonging with friends. Headphones might mean control over a noisy bus ride. When kids name the *why*, they're more willing to tolerate the *wait*.

The goal ladder: break the big want into small wins

A big goal can feel like a locked door. Kids push on it once, feel no movement, and give up-or start negotiating for a bailout. Your job is to show them the door has a handle. The handle is breaking the goal into steps.

Think of it as a ladder:

- The top rung is the full purchase.
- The middle rungs are smaller checkpoints ("I'm halfway!").
- The bottom rung is the first deposit.

Checkpoints are not childish. Adults use them constantly: progress bars, fitness streaks, mortgage amortization charts, retirement milestones. The brain likes evidence that effort is working.

One practical way to create rungs is to pick a handful of round-number milestones: 25%, 50%, 75%, and 100%. When a child hits 25%, celebrate briefly-nothing huge, just a moment of recognition. "You're a quarter of the way there." That statement is a tiny dopamine reward that doesn't require spending money.

The simple timeline formula that changes everything

Kids tend to ask, "Can I get it?" A planner asks, "When can I get it?" That shift is the heart of goal setting.

Here's the core formula, written in kid-friendly language:

$$\textit{Time to reach goal} = \textit{Goal cost} \div \textit{Money saved per week}$$

It's intentionally plain. No interest rates, no complicated assumptions. Just arithmetic that turns a wish into a calendar date.

To use it, you need two numbers:

- *Goal cost:* the real total, including tax, shipping, accessories, or anything else that must be purchased with it.
- *Money saved per week:* not money received, but money actually set aside.

That second point is where kids get tripped up. They hear "I get $10 a week" and assume they save $10 a week. But if they spend $6 and save $4, then $4 is the number that matters. The plan isn't what you earn; it's what you keep.

If your child uses jars or buckets, you can make this very concrete: "How much goes into the Save bucket each week?" That's your savings rate—the percentage of incoming money that gets saved. (Savings rate is a grown-up term; kids can just call it "my save amount.")

A concrete case: Jayden and the $90 sneakers

Jayden is 11. He wants sneakers that cost $90. Not "nice sneakers." *These* sneakers—because his friend has them and because the internet has informed him that this is what cool looks like.

Jayden gets $8 a week in allowance. He also has an "extra jobs" list he can choose from, like washing the car for $10 or pulling weeds for $6.

If you ask Jayden, "Do you want to save for sneakers?" he'll say yes and then spend his money on something else by Friday. So his parent does the planning conversation at the table with a piece of paper and a pencil (or a notes app). They keep it calm and factual, like planning a trip.

Step 1: Get the real cost.

- Sneakers: $90
- Sales tax (estimated): $7
- Total goal: $97

Step 2: Decide the weekly save amount. Jayden wants to still have some spending freedom, so he chooses to save $5 per week from allowance.

Step 3: Calculate the timeline.

$$97 \div 5 = 19.4$$

That's about 20 weeks. Jayden's face does the thing faces do when they meet reality.

Now comes the most important parenting move in goal setting: you don't argue with the face. You offer levers.

Lever 1: Increase money in. Jayden decides to do one paid job every other week for $10.

Lever 2: Increase money saved. Jayden can choose to save $6 per week instead of $5.

They test a new plan:

- Save from allowance: $6/week
- Extra jobs: $10 every other week—approximately $5 per week averaged
- Total saved per week: $11/week

Timeline:

$$97 \div 11 \approx 8.8$$

Call it 9 weeks.

Jayden can handle 9 weeks. It's not fun, but it's within his emotional range. And now the sneakers have transformed from a source of whining into a project with steps. He writes "Week 1: $11" on paper. He starts.

Week 4, Jayden is invited to a birthday party and wants to buy a gift. He's annoyed because his spending money is lower now. His parent doesn't rescue him. They ask, "What do you want more—sneakers sooner or more spending right now?" Jayden grumbles and chooses the sneakers. That grumble is not failure. That grumble is *a trade-off being learned in real time.*

The underrated part: adding "hidden costs" so your child doesn't feel tricked later

One reason kids abandon goals is the surprise of extra costs. The item price is not the total. Adults know this. Kids don't—yet.

Teach "all-in cost" early:

- tax,
- shipping,
- accessories (a case, a charger, extra game controller),
- maintenance (bike lock, helmet).

You can say, "We're not doing this to make it harder. We're doing it so you don't reach the finish line and discover it moved."

This is a quiet but powerful money lesson: the world loves to advertise the smallest number.

Implementation intentions: the magic of an if---then plan

A plan like "I'm going to save more" is an intention. Intentions are fragile. They disappear when the day gets busy, when a friend texts, when the store has a sale.

There's a more durable kind of plan called an *if—then plan*. It sounds like this:

> *If* I get money, *then* I immediately put $X into my Save bucket.

Or:

> *If* I'm tempted to buy something, *then* I wait 24 hours and check my goal picture again.

This works because it removes decision-making in the moment of temptation. It pre-decides. Pre-deciding is how adults automate good behavior: paycheck deductions, automatic transfers, calendar reminders. Kids can do a version of it too, and they often love it because it feels like a "hack."

If your child is younger, make it concrete:

> "If money touches your hand, it touches the jars next."

If your teen uses a bank account:

> "If money hits checking, I transfer $X to my goal bucket the same day."

Keeping goals flexible without making them meaningless

Kids change their minds. Sometimes because they're fickle, sometimes because they're learning what they actually value. A goal system that treats changing your mind as a moral failure will backfire. You want flexibility with structure.

Try a simple "cooling-off rule" for switching goals:

> "You can change your goal, but you have to wait one week and write down why you're switching."

That pause separates a real preference change from a momentary spike of desire. It also teaches reflection: "Is this new thing actually better, or just new?"

If your child switches goals after the week, let them—without sarcasm. Goal setting isn't about forcing one purchase. It's about practicing planning. Planning is the win.

The parent's role: coach, not financier

Parents often become the "gap filler": "You're $18 short? I'll cover it." It's understandable. You want your kid to be happy, and you don't want a long wait to turn into daily complaining.

But bailing out shortfalls regularly teaches a brutal lesson: planning is pretend, because the bank of Mom and Dad is real.

A better role is the coach with guardrails:

- Help define the goal and all-in cost.
- Help choose a savings amount that's challenging but realistic.
- Help build an if—then plan for consistency.
- Then step back and let time do its work.

If you want to add motivation without undermining the process, use *matching* instead of rescuing. Matching is when you contribute a small amount based on their saving behavior, not their complaining. For example: "For every $10 you save, I'll add $1." Now you're rewarding the habit, not the desire. (And you can set limits so it doesn't become open-ended.)

Teaching patience without teaching deprivation

There's a fear parents don't always name: "If I make my kid wait, am I being harsh?" Waiting isn't harsh. Unclear waiting is harsh. Endless waiting is harsh. Waiting without agency is harsh.

Goal setting is the opposite. It gives agency:

- You can get it faster by saving more.
- You can get it faster by earning more.
- You can choose a cheaper version.
- You can decide it's not worth it.

That's not deprivation. That's power.

Actionable takeaway: the 10-minute Goal Plan worksheet (no printer required)

Tonight, pick one thing your child wants and do this on a piece of paper:

1. *Name it:* Write the goal and tape/draw a picture.
2. *Price it:* Write the all-in cost (item + tax + extras).
3. *Fund it:* Write how much will be saved each week (from allowance/earnings).
4. *Date it:* Divide cost by weekly savings and circle the number of weeks.
5. *Lock it in:* Write one if—then rule: "If money comes in, then $X goes to my goal bucket immediately."

Then put the paper somewhere your child can see it. Not to pressure them—so the future stops being fog and starts being a plan they can touch. The moment your child says, "I'll have it in seven weeks," they've moved from wishing to building, and that's a skill that keeps paying long after the sneakers wear out.

3.6 Building Resilience: The Emergency Fund

The first "emergency" your child experiences is rarely cinematic. It's usually something small and sticky: a lost library book, a cracked phone screen protector, a forgotten lunch, a band trip permission slip that somehow needs cash *tomorrow*. But the feeling is big. The feeling is that hot little wave of *uh-oh*—and what kids learn in that moment often becomes their default money reflex for years.

Some kids learn: *When I mess up, someone rescues me.* Some kids learn: *When I mess up, I'm in trouble.* The emergency fund teaches a third option: *When something goes wrong, I have a plan.*

Adults call it an emergency fund. For kids, it's easier and friendlier to call it a *buffer*. A buffer is a soft edge. It's the difference between a bump and a crash. And it's one of the most emotionally powerful financial tools you can give a child, because it turns money into safety instead of drama.

Why kids need ``emergency money" even if parents can cover it

A common parent thought is: "My child doesn't need an emergency fund. I'm the emergency fund." True—until it quietly isn't.

If you always cover surprises, your child never has to plan for reality. They learn that money problems are solved by asking an adult, not by anticipating, saving, and recovering. That's not a character flaw. That's just the environment you built.

If you never cover surprises, your child can learn something equally unhelpful: that mistakes are catastrophes and money is a constant threat. That's how you get a kid who panics, hides problems, or lies to avoid consequences.

A buffer fund is a middle path that trains the nervous system as much as it trains the math. It says: "Surprises happen. We expect them. We prepare. We rebuild." That cycle is the essence of financial resilience.

In the real world, plenty of adults don't have that cycle. Many households report they would struggle to cover a relatively modest emergency expense without borrowing or scrambling. If you want your child to grow into someone who can handle life's inevitable little shocks, practice has to start when the shocks are tiny and safe.

What counts as a ``kid emergency"? (Hint: not everything inconvenient)

Kids will happily classify many things as emergencies:

- "My friend's birthday party is in two hours and I need a gift."
- "The snack bar is open and I have no money."
- "Everyone is buying this thing today."

Those are urgent wants, not emergencies.

A kid emergency is something that is either (1) a genuine mistake with a real consequence, or (2) an unavoidable surprise that disrupts the plan. Examples:

- Lost library book (fee is real).
- Missed the bus and needs a ride (not a daily habit, but a one-off).
- Sports mouthguard broke; needs replacement to play.
- Forgot lunch money on a day when there isn't a backup option.

The secret is that emergencies should feel slightly annoying, not thrilling. If the buffer fund becomes "fun money with a heroic label," it will be drained constantly. A good buffer has rules, and the rules are calm.

The buffer fund mindset: expect the unexpected (without becoming anxious)

There's a way to teach "be prepared" that accidentally teaches fear. The goal isn't to make kids worry. The goal is to normalize reality.

> "Life has surprises. We don't panic about them; we plan for them."

When kids internalize that, they start to see financial preparedness as competence, not as doom.

How big should a kid's emergency fund be?

Adults often measure emergency funds in months of expenses. Kids don't need that. They need a buffer that matches their world.

A useful starting point is a *small, capped* buffer:

- Ages 6–9: $10–$20
- Ages 10–13: $20–$50
- Ages 14–18: $50–$150 (depending on responsibilities and what they pay for)

The cap matters because it creates a clear finish line. Kids are more willing to save for a buffer when they can see it being "done." Once it hits the cap, you stop contributing to the buffer and redirect that money to goals, giving, or long-term saving. That prevents the buffer from feeling like an endless pit that steals from fun.

Where the buffer lives: a fourth jar, a hidden envelope, or a teen bucket

If your child uses the three-jar system, the buffer can be a *fourth container*: *Buffer*. Put it slightly out of reach and treat it differently. The message is: "This is not your everyday money."

If your child is older and uses buckets, the buffer is its own bucket: *Emergency/Buffer*. If your teen has a bank account, you can treat the buffer like a mini savings account earmarked for surprises.

The crucial detail is labeling. The label changes behavior. A pile of money is tempting. Money with a job is disciplined.

Funding the buffer: small and steady beats dramatic and rare

Kids often want to fill the buffer instantly so they can stop thinking about it. Adults do the same thing: "When I get my bonus, I'll start saving." The problem is that "someday" funding rarely arrives on time.

Better: tiny contributions that happen automatically. A clean rule is:

> "Until the buffer is full, $1 (or $2) from every money event goes into it first."

This works with allowance, birthday money, and earnings. It's the kid version of paying yourself first. It also keeps the buffer from competing emotionally with a child's main savings goal. One dollar feels manageable. Ten dollars feels like a moral argument.

If you want to add a parent boost, consider a match:

> "For every dollar you put in your buffer, I'll add a dollar until it hits the cap."

Matching changes the emotional tone from "taking away my money" to "building safety together."

A real-world case study: the lost trumpet mouthpiece

Carla is 14 and plays trumpet in the school band. One morning she realizes her mouthpiece is missing. She's sure it was in her case. She panics because rehearsal is that afternoon and the band director is strict about being prepared.

In many families, this becomes a heat moment: accusations, frantic searching, and then the parent solution-either buying a replacement immediately with resentment, or refusing and letting the child suffer in a way that feels punitive.

Carla's family tries a buffer system.

Carla has a $75 buffer bucket in her bank app. It took a few months to build because $5 from each babysitting job went into the buffer until it was full. Her parents also made one rule: buffer money can be used for *repairing real-life consequences*, not for "I want it now."

Carla buys a replacement mouthpiece for $18. She feels the sting-because it's her money-but she also feels something new: *relief without begging*. She goes to rehearsal prepared.

Then the next part happens, and it's the part that actually builds resilience: her buffer is now $57, not $75. It's not "broken." It's simply lower. Carla creates a refill plan: $3 from each week's allowance, plus $5 from her next babysitting job, until she's back at $75.

This is the cycle you want your child to learn:

Emergency happens → use buffer → stress stays low → refill buffer.

Notice what's missing: shame. The lesson isn't "how could you." The lesson is "this is what buffers are for."

Two weeks later, they find the original mouthpiece under the passenger seat of the car. Carla laughs, mildly horrified. Her parent says, "Great—now you have a backup." The surprise becomes a story, not a trauma.

Teaching the difference between ``rescue" and ``support"

The buffer fund changes your role as a parent. You don't disappear. You become a calmer guide.

When a kid emergency happens, your job is to help them label it and choose the right tool:

- Is this a true emergency or an urgent want?
- If it's a true emergency, does the buffer cover it?
- If the buffer doesn't cover it, what's the plan—borrow from a goal fund, earn extra, or accept the consequence?

This is support. Rescue is when you remove the consequence *and* remove the learning. Support keeps dignity intact while keeping reality intact.

A phrase that helps keep you out of the rescue habit:

> "I can help you solve this, but I won't solve it *for* you."

What to do if your child empties the buffer too often

If the buffer is constantly being used, one of two things is happening:

1) *The rules are too loose.* Your child is treating it like a second Spend bucket. Tighten the definition of emergency and add a pause:

> "We wait 10 minutes before using buffer money unless it's truly time-sensitive."

That pause is often enough to downgrade a "crisis" into a want.

2) *Your child is disorganized, not irresponsible.* Executive function (planning, remembering, tracking) develops over time, and some kids need more scaffolding. If the emergencies are mostly forgotten items, build systems: packing checklist by the door, a weekly "reset," a designated spot for band gear. The buffer shouldn't be a penalty for a brain that's still under construction.

If you notice a pattern-lost water bottle every week-consider a hybrid approach: the first replacement comes from the buffer, the second comes from the child's Spend money, and the third triggers a system change (labeling, attaching, routine) before another purchase. The goal is not to punish. It's to reduce repeat emergencies.

How the buffer fund teaches emotional regulation

One of the sneakiest benefits of a buffer fund is that it teaches kids to regulate emotions around money. Money problems feel personal. They feel like failure. The buffer fund reframes them as events: "This happened. We handle it."

That skill matters in adulthood more than we like to admit. Financial trouble isn't only about math; it's about panic. A person who has practiced "handle, then refill" is less likely to spiral into avoidance, denial, or impulsive decisions. They're more likely to take the boring, powerful step: making a plan.

And that's what resilience is—boring competence under pressure.

Actionable takeaway: build a kid buffer in two weeks with a cap and a refill rule

Tonight, decide three things and write them down with your child:

1. *Buffer cap:* Choose a number that fits your child's age (e.g., $20, $50, $75).
2. *Funding rule:* "Until the buffer hits the cap, $2 from every allowance/earning goes into the buffer first."
3. *Refill rule:* "Any time we use buffer money, we refill it before we add extra to Spend."

Then pick a place for the buffer to live—fourth jar, labeled envelope, or a bank bucket—and label it clearly. The label is a promise: when something goes wrong, your child won't have to panic or perform. They'll have a tool, a process, and the steady confidence that the problem is solvable—because they've solved it before.

Chapter 4

Spending Wisely and Giving Back

If saving is the vegetable of personal finance, spending is the dessert-everyone wants it, but overindulgence leads to a stomachache. In this chapter, we reframe spending not as the enemy of wealth, but as a tool for living a life aligned with one's values. We will move beyond the restrictive 'don't buy that' mindset to build a proactive framework for decision-making. You'll learn how to help your children differentiate between fleeting wants and genuine needs, navigate the psychological minefield of social media influence, and realize that the most satisfying purchase is often one made for someone else.

4.1 Budgeting Without Tears

The word *budget* has a public-relations problem.

Say it at the dinner table and you can almost hear the shutters come down: *No. Less. Can't. Stop.* Kids pick up that vibe fast. If the first lesson they learn is that money planning equals punishment, they'll treat it the way most adults treat punishment: avoid it until something breaks.

So change the name and, more importantly, change the emotional job of the tool. A *Spending Plan* isn't a cage. It's a permission slip. It's the family's way of saying: *Yes, you can spend on things you love—as long as you decide on purpose.*

That one shift—*from restriction to intention*—is how you teach budgeting without tears. Not by perfect spreadsheets. Not by scolding. By giving your child a simple map and letting them practice steering.

A spending plan is a story about choices

A good spending plan answers three kid-sized questions:

- *What is my money for?*
- *What do I do when I want two things at the same time?*
- *What happens when I mess up?*

Those questions sound basic, but they're the same questions adults wrestle with when they're deciding between takeout and groceries, between a nicer car and an earlier retirement, between donating and upgrading their phone. The difference is that adults have bigger numbers and more consequences.

Kids need practice with consequences that sting a little—but don't scar.

Here's the hidden win: a spending plan teaches *trade-offs* without turning you into the villain. When the plan says, "You spent your Fun money already," you're not saying "no." Reality is saying "not unless you move money from somewhere else." It turns a parent-child power struggle into a math problem a kid can solve.

The ``three-bucket" system that doesn't feel like homework

For most families, the simplest useful structure is three categories:

- *Fun*: immediate joy (snacks, games, small toys, going out with friends)
- *Goals*: future joy (saving for something bigger)
- *Giving*: other people's joy (donations, gifts, helping out)

You can start these categories as soon as your child is old enough to understand "now" versus "later." If they're very young, do it with literal jars. If they're older, do it with a notes app, a simple table on paper, or three balances in a basic bank account (some banks allow multiple "savings goals" buckets; if yours doesn't, you can track it manually).

The categories aren't magic; they're memorable. What makes them work is that *each bucket gets a job*. Money without a job is snack money. Money with a job is freedom.

A common parent worry is, "If I let my kid decide, they'll put everything in Fun." Sometimes, yes—especially at first. That's not failure; that's data. You're not trying to produce a child who is naturally frugal at age nine. You're building the muscles of noticing, choosing, and adjusting.

Start with a plan so small it's hard to mess up

Adults often hand kids a money system that's too big on day one: multiple categories, rules, percentages, a lecture about compound interest, and a look of disappointment when the kid immediately buys slime.

Instead, start with an amount and a cadence that match your child's world.

Weekly works better than monthly for most kids because their time horizon is shorter. A week is something they can feel. A month is a foggy eternity.

If your child gets $10 a week, you might begin with:

- Fun: $6
- Goals: $3
- Giving: $1

If they earn money irregularly (birthday cash, occasional chores), you can still do a "split" each time money arrives.

Don't be afraid to set a default split for younger kids. The plan is not about squeezing them; it's about giving them a stable starting point. As they get older, you can let them propose the split—and justify it.

That justification is where the learning lives.

Make the plan visible, or it becomes imaginary

A spending plan only works if a kid can *see* it. Otherwise it turns into a vague parental vibe: "You've spent a lot lately." Vibes don't teach. Numbers teach.

Choose a simple format and keep it consistent. Here's a paper version that works for ages roughly 8–14, and still feels respectful to teens:

Date	*What happened*	*Fun*	*Goals*	*Giving*
Feb 1	Allowance / pay	+6	+3	+1
Feb 3	Snack at school	-2		
Feb 6	Saved toward headphones		-3	
Feb 8	Donation to animal shelter jar			-1

This looks boring to adults who love apps, but it's powerful for kids because it makes money concrete. Every line is a tiny cause-and-effect lesson.

If your child is older and wants digital, use whatever tool they'll actually use. The best spending plan is the one they won't avoid. A notes app table, a simple spreadsheet, or a kid-friendly debit app can all work. Your job is to keep it from becoming a black box. If the app hides the "why," you add it back: "What bucket did that come from? Was it worth it?"

The secret ingredient: ``Yes, if"

The most parent-saving phrase in budgeting is "Yes, if."

- "Can I buy this?"
  "Yes, if it's in your Fun bucket."
- "But I really want it."
  "Yes, if you're willing to move money from Goals to Fun."
- "But I don't want to."
  "Then it's a 'not yet."

This is not a debate about whether you approve of the purchase. It's practice in managing constraints.

Kids often hear "no" as "you don't get me." "Yes, if" lands differently: "I do get you—and you still have to choose."

Over time, your child starts to anticipate the question you'll ask. That's when you know the tool is working: they're internalizing the trade-off.

A real-world case study: the hoodie, the game, and the field trip

Consider Maya, age 12. She gets $15 a week: $9 Fun, $5 Goals, $1 Giving. She's saving for a $60 hoodie everyone at school seems to have. The hoodie isn't a need, but it feels like one, because middle school is a social ecosystem and nobody wants to look like an outsider.

Week 1: Maya puts her $5 Goals away and feels proud. Then a new mobile game offers a "limited time" bundle for $8. She buys it. It's exciting for a day, then the excitement evaporates. She doesn't tell you because she already knows how the conversation goes: disappointment, lecture, gloom.

Here's where a traditional budget tends to backfire. If the plan is framed as "You broke the rules," Maya's brain learns one lesson: *Hide the evidence.*

A spending plan framed as "You're the decision-maker" teaches a better lesson: *Own the decision. Learn from it.*

So you do a five-minute review together at the end of the week, not a courtroom drama. You keep your face calm and curious.

"Where did the Fun money go?"

She tells you. You nod.

"Did it do the job you hoped it would do?"

She shrugs. "Not really."

"Okay. What do you want your money to do next week?"

Now she's thinking like a planner, not a defendant. She decides to cap game spending at $2 a week so she can still buy snacks with friends and keep saving for the hoodie.

Week 3 brings the twist: a school field trip fee due Friday, $12. You had forgotten about it; she definitely forgot about it.

This is where families either spiral into blame or teach competence.

If you can afford it, you might cover the fee but require a *repayment plan* from her Fun bucket over the next two weeks. If you can't cover it, you work with her to problem-solve: can she do an extra chore? Can she use a portion of her Goals money and adjust her hoodie timeline? Can she borrow from you and pay back with a small "interest" amount (like an extra $1) to teach the cost of borrowing?

The key is that the spending plan gives you options that aren't moralizing. The plan doesn't say "you're bad with money." It says: *This happened. Now decide what to trade.*

And that is exactly what adulthood demands, except with rent and car insurance instead of hoodies and field trips.

How to do the weekly review without making it weird

A spending plan review should feel more like checking a scoreboard than receiving a report card. Fast, neutral, routine.

A simple script:

- *What came in?* (allowance, pay, gift)
- *What went out?* (two or three biggest items)
- *Any surprises?* (fees, forgotten commitments)
- *Any regrets?* (no shame—just noticing)
- *One tweak for next week?* (a small adjustment)

Aim for 5–10 minutes. Stop while it's still painless.

If you sense your child bracing for criticism, you can say out loud: *"This isn't a 'gotcha.' This is practice."* Kids who feel safe telling the truth learn faster. Kids who feel judged learn to curate.

Overspending: teach resilience, not guilt

At some point your child will overspend a category. Not might. Will.

The teaching moment isn't "How could you?" The teaching moment is "What now?"

When the Fun bucket hits zero on Wednesday, you have three broad options:

1. *Do nothing*: They wait until next week. This teaches patience and consequence. It's surprisingly powerful.
2. *Allow a transfer*: They move money from Goals to Fun. This teaches trade-offs and opportunity cost (choosing one thing means delaying another).
3. *Offer a way to earn*: An extra chore, a small family job, or selling something they no longer use. This teaches

agency, but use it carefully so "oops" doesn't become a strategy.

Pick the option that matches the situation and your family values. If the overspending was impulsive and frequent, "do nothing" is often the cleanest teacher. If the overspending came from a genuine surprise (school fee, broken headphones), a transfer or earn option can teach flexibility without making them feel helpless.

One caution: avoid "bailouts" that have no repayment or tradeoff. The lesson your child's brain quietly files away is: *Buckets are pretend; parents are the real ATM.* A spending plan only builds capability if the boundaries mean something.

Needs, wants, and the third category nobody talks about: values

Most families teach "needs vs. wants" as a binary: needs are responsible, wants are frivolous. Kids can smell the judgment on that a mile away.

A more useful framing is:

- *Needs*: things required to live and function (food, basic clothing, school supplies)
- *Wants*: things that are nice to have (treats, entertainment, extra gear)
- *Values*: the reasons you choose certain wants on purpose (creativity, friendship, health, learning, generosity)

A soccer ball is a want. If your child values movement and teamwork, it's a want with a job. A fancy water bottle is a want. If it's primarily about status, it's a want with a different job. Neither is automatically "bad," but naming the job turns mindless spending into conscious spending.

You can ask, gently and often: *"What value does this support?"* Not as a trap. As a flashlight.

When kids learn to connect spending to values, they stop needing you to be the external voice of discipline. They develop an internal compass.

Age-appropriate upgrades: keep the core, expand the freedom

The three buckets can grow with your child without losing the simplicity.

Ages 5–7: Physical jars, simple splits, very short reviews. The goal is "money is finite" and "saving is a real thing."

Ages 8–10: Add the idea of "planned spending." For example, if they know they want movie snacks on Saturday, they reserve money for it on Monday. That's executive function disguised as popcorn.

Ages 11–13: Let them propose their own percentages. Add one more category if it fits their life, like "Friends" or "Clothes." Keep reviews short and factual.

Ages 14–18: Introduce irregular expenses and longer horizons: sports fees, dances, gifts, subscriptions, gas. This is where a teen-friendly spending plan becomes a launchpad for adult skills: planning ahead, comparing options, and owning consequences.

Notice what stays the same: categories with jobs, visible tracking, routine reviews, and "Yes, if."

Common traps (and the fix that keeps trust intact)

Trap 1: Making the plan a surveillance tool.
If every line item turns into interrogation, your child will stop recording purchases. The fix is to reserve "deep dives" for truly important patterns (like constant online microtransactions) and keep ordinary spending neutral.

Trap 2: Using the plan to enforce your preferences.
If your child spends Fun money on something you think is dumb, you can still let the consequence teach. Your opinion can exist without running the show: *"Not how I would spend it, but it's your Fun bucket."* This is hard. It's also how you raise a decision-maker.

Trap 3: Overcomplicating too early.
Ten categories feels "real," but it's often just a way to feel in control. The fix is to start simple and expand only when the kid's life demands it.

Trap 4: Ignoring your own spending plan.
Kids notice when adults say "We can't afford that" and then swipe a card on impulse. You don't have to be perfect; you do have to be honest. One of the most powerful sentences a parent can say is: *"I didn't plan for that. I'm going to move money from another category."* That's not weakness. That's modeling.

A spending plan should make your child feel powerful

Done well, a spending plan gives your child three kinds of power:

- *Power over timing*: "Not now" doesn't mean "never."
- *Power over trade-offs*: "I can pick this, but it pushes that."
- *Power over identity*: "My money reflects what I care about."

That's a very different emotional experience than "My parent is controlling my money."

It's also the point. The goal isn't a child who never makes a messy choice. The goal is a child who learns that a messy choice is not a moral failure—it's a moment to adjust the plan.

Actionable takeaway: the 10-minute Spending Plan kickoff

Pick a day this week and do a short kickoff. Ten minutes is enough.

1. Choose three buckets: Fun, Goals, Giving.
2. Pick a weekly amount (allowance, pay, or a portion of earnings) and a simple split.
3. Make it visible: jars, paper table on the fridge, or a shared note.
4. Schedule a weekly five-minute review at a predictable time (Sunday evening works well).
5. Practice the phrase *"Yes, if"* once—out loud—so it's ready when the first purchase request comes.

If you do nothing else, do that. A spending plan that gets used imperfectly will teach more than a budget that looks impressive and lives in a drawer.

4.2 Mastering Impulse Control

The most expensive part of an impulse purchase isn't the money. It's the *speed.*

One minute your kid is fine. The next minute they're vibrating with certainty that they *need* the limited-edition water bottle, the game skin that "won't come back," the squishy toy by the checkout, the in-app bundle with a countdown clock. The pitch is always the same: *Decide right now or lose it forever.*

That urgency isn't accidental. It's a shortcut around the part of the brain that asks boring questions like, "Do I already have something that works?" and "What else did I want my money to do this week?" When the goal is to sell, slowing down is the enemy.

So your family gets to build a tiny piece of friendly friction: a habit that turns "buy now" into "pause, then choose."

A quick, sane definition of impulse buying

Impulse buying is what happens when a want shows up wearing a costume that looks like a need. It's the purchase you make *before* you've decided if it's worth the trade-off. And it isn't a character flaw. Kids aren't "bad with money" because they feel pulled by shiny things; they're human.

A useful way to explain this to a child is: your brain has two speeds. Fast brain grabs. Slow brain plans. Fast brain is great when a soccer ball is flying at your face. It is not great when an ad is flying at your feelings.

The goal isn't to delete fast brain. The goal is to teach your kid how to *switch gears*.

The 24-Hour Rule: a speed bump that saves real money

The simplest anti-impulse tool is also the least dramatic:

If it's not a need, wait 24 hours before buying it.

That's it. No lecture. No shame. Just time.

Why it works: the emotional spike that makes a purchase feel urgent usually fades. After a day, your kid can still buy the thing if they genuinely want it—but now it's a choice, not a reflex.

A few clarifying details that keep the rule from collapsing in real life:

- *Needs* can be immediate: lunch because you forgot it, replacement calculator battery for homework, a required school fee.
- *Non-urgent wants* wait: toys, treats, cosmetics, games, merch, upgrades, "just because" buys.
- *Social-time purchases* can have a version of the rule: if friends are going out "today," your kid can decide on the spot—but only within their pre-decided spending plan category. The "wait" happens for the extra add-ons:

dessert, the souvenir, the bonus in-app purchase afterward.

This is where your earlier "Yes, if" language becomes a superpower. The rule doesn't sound like "no." It sounds like: *Yes—tomorrow, if you still want it and it fits your plan.*

Immediately after you introduce the 24-hour rule, give your kid a concrete map to follow in the moment:

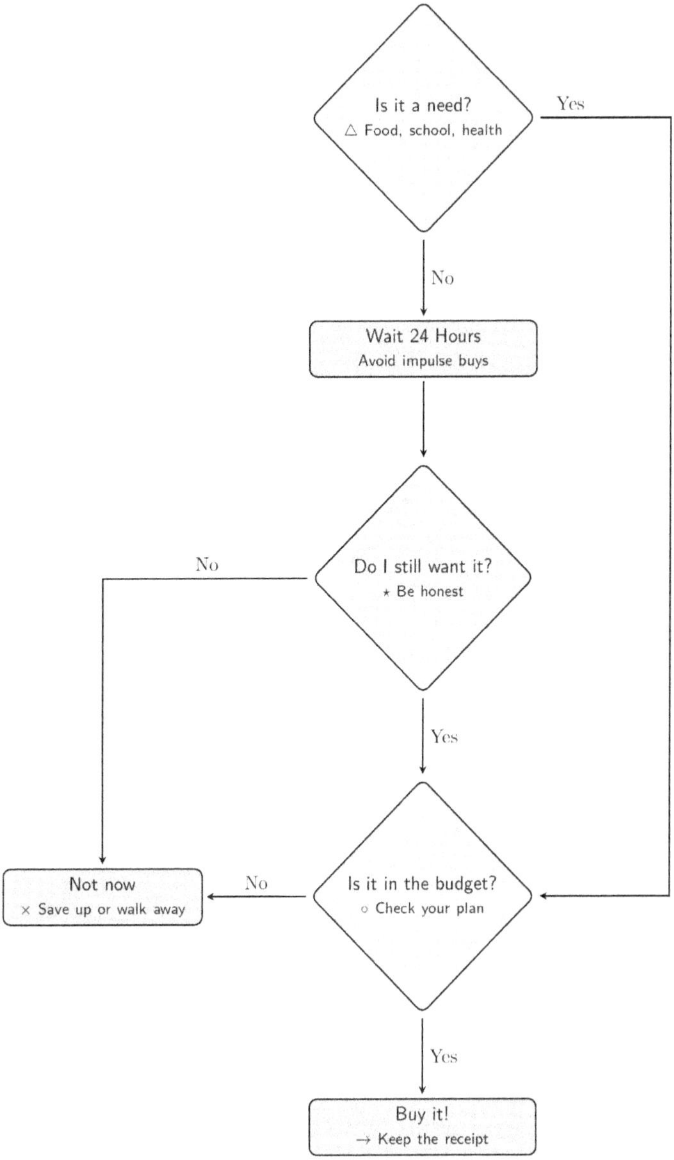
Is it a need?
△ Food, school, health
Yes
No
Wait 24 Hours
Avoid impulse buys
Do I still want it?
⋆ Be honest
No
Yes
Not now
× Save up or walk away
No
Is it in the budget?
○ Check your plan
Yes
Buy it!
→ Keep the receipt

Wish lists: a parking lot for ``I want it"

Wish lists sound like something you do for birthdays. Use them like a tool.

A wish list is a safe place to put desire so it doesn't bounce around your kid's brain all day. It's a way of saying: *I'm not ignoring what you want. I'm storing it.*

For younger kids, the wish list can be a note with drawings or printed pictures. For older kids, it can be a notes app or a shared list. The format matters less than the habit: when the urge hits, it goes somewhere other than the checkout line.

A sneaky benefit: wish lists reveal patterns. After a month, your child will notice they wanted twelve things and only remember three of them. That's not a lecture; it's evidence.

If you want to make the lesson even sharper, add one column: *"Why do I want this?"* Not as a therapist question. As a detective question. "It looks cool" is allowed. "My friend has it" is allowed. "I'm bored" is allowed. Naming the why turns a vague craving into something you can think about.

A real-world case study: the $9.99 problem

A lot of modern impulse buying doesn't look like a shopping bag. It looks like *$9.99*.

Take Jordan, age 13, who plays a popular game with friends. The game itself is free, but the store inside the game is basically a candy aisle that follows you around. New skins rotate weekly. Special "starter packs" pop up after you win. There's a bright button that says "Best Value!" next to a bundle of gems.

Jordan starts spending "little" amounts: $4.99 here, $9.99 there. It doesn't feel like spending because nothing changes in his hands—no cash leaves his wallet. Then you notice the charges.

If you respond with a speech about responsibility, Jordan hears: *"You're in trouble. Hide it better."*

If you respond with a system, Jordan learns: *"I can enjoy games and still stay in control."*

Here's what the system can look like:

- Jordan's spending plan includes a *Gaming* line inside Fun money.
- Any in-app purchase follows the 24-hour rule, even if it's "limited." Limited is the whole trick.
- Jordan keeps a wish list of game items, with prices.
- Once a week, he picks *one* item (or none) and buys it *on purpose*, within his Gaming amount.

Two things happen. First, the random spending stops. Second, the purchases Jordan does make feel better. He gets the skin he actually cares about, not the one that happened to flash at him after a match.

That's an underrated part of impulse control: it doesn't drain joy. It improves it. Planned treats hit harder than accidental ones.

Teach the feeling: ``The urge is a wave"

Kids often believe an urge is a command. "I want it" becomes "I have to."

Give them a different metaphor: an urge is a wave. It rises, crests, and falls—even if you don't buy anything. The first few times they practice letting the wave pass, it can feel uncomfortable, like holding your breath. Then they realize something that makes them feel oddly powerful: *I can feel a strong want and not obey it.*

You can coach this in real time without getting dramatic:

- "Rate your want from 1 to 10."
- "Okay, we're waiting ten minutes. Let's see what the number does."

Ten minutes is often enough for the number to drop. Not to zero, but to something manageable. For younger kids, do it while you keep shopping. For teens, do it by setting a timer and moving on.

This isn't about "being tough." It's about learning that emotions are informative, not authoritative.

Add friction where it counts: make impulse a little harder

Willpower is unreliable. Everyone's willpower is unreliable. The trick is to build an environment that doesn't require heroic self-control.

A few parent-tested friction points:

- *No stored cards on devices for kids.* If they have to ask you to type in payment details, you've created a pause.
- *Turn off one-click buying.* Convenience is great for restocking paper towels; it's terrible for late-night scrolling.

- *Use a separate "kid spend" card with limited funds.* A hard limit can prevent a bad hour from becoming a bad month.
- *Put the phone to charge outside bedrooms at night.* A tired brain makes faster, worse decisions. (Adults, too.)

These aren't punishments. They're guardrails. Nobody calls a seatbelt a punishment. It's a safety device you use because you're human and physics is real.

Talk about marketing like it's a magic trick

Kids are less embarrassed about being influenced if you frame persuasion as a *skill* someone is using on them—not a weakness inside them.

The best conversations sound like you're analyzing a magic trick together:

- "What's the ad trying to make you feel?"
- "What's the timer for? Who does the timer help?"
- "What problem is this product pretending to solve?"

Even better: pick a real example in the wild. The checkout aisle is a museum of impulse design: candy at kid-eye level, small toys, bright packaging, "only $1!" signs. Online shopping has its own museum: "Only 2 left," "People are buying this right now," auto-playing videos, influencer codes, free-shipping thresholds that nudge you to add one more thing.

When kids learn to see these tactics, they don't become immune—but they do become harder to rush.

Delayed gratification, without the cheesy moral lesson

A lot of adults heard the story that successful people can delay gratification and everyone else can't. That story is catchy—and incomplete.

What's more useful for kids is a simpler, kinder message: waiting is easier when you have a plan and you trust you'll actually get the thing later.

That's why "save for it" works better than "just don't." Saving is delayed gratification with a destination. It's not deprivation; it's scheduling.

One practical tool: *make waiting visible*. If your child is saving for a $40 item, draw a progress bar on paper and color it in each week. If they're older, track it in a notes app. The brain likes progress. Progress makes waiting feel like action, not helplessness.

How to respond when your kid slips

They will slip. Especially at first. Especially when they're tired, hungry, stressed, or trying to fit in. The response you choose determines whether the slip becomes a lesson or a secret.

A useful script:

- "That happened."
- "What did you feel right before you bought it?"
- "What did you hope the purchase would do for you?"
- "Did it work?"
- "What's the plan next time the feeling shows up?"

Notice what's missing: "What were you thinking?" That phrase almost always lands as "You're stupid." You can keep your standards high without making your child feel small.

Then add a consequence that teaches, not humiliates. If they blew their Fun money early, they wait. If they spent in a way that broke a family rule (like buying something online without permission), the consequence might be tighter friction for a while: no stored cards, extra check-ins, or a temporary pause on in-app purchases. The logic is: *We're changing the system so this is easier next time.*

Actionable takeaway: the ``Pause Kit" you can set up tonight

Before the next shopping trip, build a tiny Pause Kit with your child:

1. Agree on your 24-hour rule for non-needs, and write it on a sticky note or in a shared note.
2. Start a Wish List with three columns: Item, Price, Why I want it.
3. Pick one friction point you'll add this week (no stored cards, no one-click, or a limited-balance spend card).
4. Choose a simple check-in time once a week: "Any wish list items still calling your name?"

Impulse control isn't a personality trait your kid either has or doesn't. It's a set of small pauses that, stacked together, turn into a life skill: the ability to want something without being owned by the wanting.

4.3 The Savvy Shopper: Price and Quality

A price tag is a tiny piece of information dressed up like a full answer.

Kids learn early to compare numbers: $19.99 is less than $29.99, so the first one "wins." That's not foolish-it's math. But it's incomplete math, the kind that works for trading baseball cards and completely fails when the choice comes with hidden extras: batteries you didn't know you needed, a subscription you have to keep paying, a charger sold separately, a pair of shoes that disintegrates right as the season starts.

If you want to raise a kid who can handle the modern marketplace-where every product comes with a menu of add-ons and every checkout screen is a gentle ambush-teach them the difference between *cheap* and *good value*. Cheap is about the sticker price. Value is about what you get for what you pay, over time, including the annoying stuff nobody puts in bold print.

The best part: kids actually enjoy learning this. "Savvy shopping" feels less like a lecture and more like becoming a detective with permission to be skeptical.

Price is what you pay. Value is what you keep.

Here's a kid-friendly definition that holds up all the way into adulthood:

> *Price is the number you hand over today. Value is the usefulness you get back, after the excitement wears off.*

The excitement wearing off is not a small detail. A lot of spending mistakes come from confusing "This feels amazing right now" with "This will still feel worth it next week." The 24-hour rule helps with the timing. Savvy shopping helps with the evaluation.

A useful mental shift for kids is to stop asking, "Is it expensive?" and start asking, *"Compared to what?"* Compared to another brand, another size, another material, another option that lasts longer, another model that uses less energy, another version that doesn't require a subscription.

That question—*Compared to what?*—is how you raise a shopper who doesn't get played.

Unit price: the cheat code hiding in plain sight

If you teach only one shopping skill, teach unit pricing.

Unit price is the cost per unit of measure: per ounce, per pound, per count, per sheet, per wash. It turns confusing packages into an apples-to-apples comparison. It's the reason a "bigger" box isn't always a better deal, and the reason a sale sign isn't proof of anything.

Kids love unit pricing because it feels like cracking a code. The shelf might say:

- Cereal Box A: \$4.49 for 12 oz
- Cereal Box B: \$5.79 for 18 oz

A kid might see "\$4.49 is cheaper." A savvy shopper sees:

$$\text{Unit price A} = \frac{\$4.49}{12} \approx \$0.37 \text{ per oz}$$

$$\text{Unit price B} = \frac{\$5.79}{18} \approx \$0.32 \text{ per oz}$$

So the bigger box is cheaper *per ounce* even though it costs more today. That's a real-life lesson in trade-offs: if you have the money now and you'll actually eat it before it goes stale, B is the better value.

If your child is younger, skip the math and use a quick rule: look for the smaller number next to "per oz" or "per count" on the shelf tag. If the store doesn't show unit prices clearly, you can do the math together once or twice and your child will start seeing the logic everywhere.

One important nuance worth teaching: the best unit price isn't always the best choice. If buying the giant container causes waste—or is more than your child can afford from their spending plan—then it's not a deal, it's clutter. Value includes *fit*.

Shrinkflation: the sneaky reason your "usual" costs more

Many kids are convinced prices rise because stores "feel like it." They don't naturally notice something subtler: sometimes the price stays the same, but the package gets smaller. Less cereal in the same size box. Fewer ounces in the same bottle shape. Fewer chips in the same bag with the same loud crinkle.

This is where unit pricing becomes more than math—it becomes protection. If your kid learns to check unit price, they can spot when their "usual" brand quietly became a worse deal. They don't have to be angry about it. They just have to be awake.

A fun family challenge in the grocery store: find one product where the unit price makes the "sale" look silly. Kids love catching the trick.

Cost per use: turning "cheap" into a long-term number

Unit price helps when you're comparing sizes. Cost per use helps when you're comparing *quality*.

Cost per use is exactly what it sounds like:

$$\text{Cost per use} = \frac{\text{Total cost}}{\text{Number of uses}}$$

"Total cost" isn't just the purchase price. It can include repairs, replacement parts, accessories, or anything you must buy for the item to do its job.

This idea changes the way a kid sees "bargains." A $20 item that breaks after two uses costs $10 per use. A $50 item that lasts 50 uses costs $1 per use. Suddenly the "expensive" one is the frugal one.

The trick is to help kids estimate uses without turning it into an engineering project. You don't need perfect numbers; you need reasonable ones. Ask:

- "How many times do you think you'll wear it?"
- "How many school days are left?"
- "How long did the last one last?"

You're teaching them to think in *lifespan*, not just price.

Case study: the headphones that ate the savings plan

A real-world example makes this lesson stick.

Imagine your teen, Sam, wants headphones. Two options:

- *Option A:* $19.99 wired earbuds from a bargain rack
- *Option B:* $49.99 mid-range wired earbuds from a reputable brand with a warranty

Sam's brain does what brains do: *Get the cheaper ones. More money left for fun.*

You let Sam choose Option A. A month later, one earbud goes quiet. Then the other one starts cutting out unless the cable is held at a weird angle. Sam buys another set. Then another. In three months, Sam has spent about $60 and has a little graveyard of tangled cords in a drawer.

Now it's not just about money; it's about frustration. Sam feels unlucky. You feel tempted to say, "I told you so."

Instead, you turn it into a shopper lesson. You sit down with the receipts (or bank app history) and do the calm math together:

- Total spent on cheap earbuds: $\$19.99 \times 3 = \$59.97$
- Total working earbuds right now: none
- Hours of annoyance: many

Then you compare it to Option B:

- $49.99 once
- Likely longer lifespan

- Warranty replacement if defective

The point isn't that cheap is always bad. The point is that cheap can be a trap when the product breaks, needs replacing, or makes the user miserable. Sam learns a powerful adult skill: sometimes paying more once is cheaper than paying less repeatedly.

If you want to go one step further, introduce a family phrase: *"The durability trap."* It gives a name to the pattern. Once a pattern has a name, kids start spotting it everywhere—shoes, backpacks, chargers, sports gear, even school supplies.

Hidden costs: the ``yeah, but..." questions

Modern products rarely arrive alone. They arrive with a parade of "yeah, but..." costs.

Teach your child to ask these questions before they buy:

- "Yeah, but does it need batteries?"
- "Yeah, but do I need a case, a screen protector, or special refills?"
- "Yeah, but does it require an app or subscription?"
- "Yeah, but do I have the right charger or adapter?"
- "Yeah, but what happens if it breaks—can it be fixed?"

These questions are a kind of financial street smarts. They prevent the classic kid experience: spending all their money on the main item and then needing you to pay for the part that makes it work. (And to be fair, adults fall for this too—printers are basically a long-running prank on humanity.)

A subscription is the sneakiest hidden cost because it turns a one-time purchase into an ongoing commitment. Some subscriptions are worth it. But a kid should learn to see the pattern: a low entry price can be bait.

A simple teaching moment: if a free trial requires a card, treat it as spending. Your teen should know the renewal date and the monthly price before they click "start free trial." If they can't explain it, they're not ready to agree to it.

Reviews: teaching kids to read the internet like a grown-up

Kids trust ratings because numbers look authoritative. A 4.7-star product feels like it has been approved by the universe.

The goal isn't to turn your kid into a cynic; it's to turn them into a careful reader. Give them a few fun, memorable checks:

- *Look for specific details.* Real reviews mention fit, material, time, and context. Fake-y reviews gush without saying anything.
- *Scan the one-star reviews.* Not to be scared—just to see what can go wrong. If the same complaint repeats, believe the pattern.
- *Notice timing.* A product that suddenly got 2,000 reviews in one week might be getting "help."
- *Watch for mismatched photos.* If the review photos look like a different product, something's off.

This skill spills into life beyond shopping. Kids who can evaluate reviews can evaluate claims. They learn that popularity is not proof, and that confidence is not the same as evidence.

Energy and operating cost: the price tag isn't the whole bill

For big purchases in a household-appliances, electronics, anything that plugs in-the sticker price is only the beginning. The rest of the cost shows up quietly on the electricity bill.

This is a great way to teach kids "total cost over time" without scaring them with adult finance. If your child is old enough to care about buying a mini fridge for their room, a gaming console, or a space heater, they're old enough to hear: "Some things are cheap to buy and expensive to run."

Many appliances come with an energy label that estimates annual energy use or cost and makes it easier to compare models. The numbers aren't perfect predictions, but they teach the right instinct: when two options do the same job, the one that sips energy can be the better deal, even if the purchase price is a little higher.

A kid-friendly analogy: buying an appliance is like buying a pet. The adoption fee is not the whole cost. You're also paying for food and care.

When "best value" is actually "best fit"

A savvy shopper isn't someone who always buys the most durable option or the biggest size. It's someone who buys what fits their real life.

Sometimes the best value is the cheaper option because:

- your child is growing quickly (expensive shoes that last two years won't last two years)

- they're experimenting with a hobby (start modestly, upgrade if they stick with it)
- the item will be used rarely (renting or borrowing might beat buying)

This is where you can teach a mature, flexible rule: *Spend more on what you use often and what matters to you; spend less on what you won't use much.* It respects your child's values and protects their money.

You can also model a powerful adult move: saying no to an upgrade because the current thing is "good enough." "Good enough" is not settling. It's choosing to put money somewhere else on purpose.

A low-drama way to practice: the ``Detective Shop" game

If you want a practical exercise that doesn't feel like schoolwork, try this on a normal shopping trip:

Pick one item your family already buys (cereal, yogurt, paper towels, soap). Give your child a small mission:

> *Find the best value option that we would actually use. You have five minutes.*

They can compare unit prices, packaging sizes, and quality cues. Then they present their choice like a detective explaining a case: "This one costs more, but it's cheaper per ounce," or "This one is cheapest per ounce, but we won't finish it before it expires, so it's not actually a deal."

You're not just teaching shopping. You're teaching decision-making with evidence-an adult superpower disguised as groceries.

Actionable takeaway: the two-question checkout pause

Before your child buys any non-need item this week, have them answer two questions out loud:

1. *What's the real total cost?* (extras, refills, subscription, accessories, repairs)
2. *What's the value math?* (unit price or cost per use-whichever fits)

If they can answer those, they're not shopping blindly. They're shopping like someone who expects their money to do a job—and refuses to be tricked by a simple number on a tag.

4.4 Navigating Social Pressure and Ads

Your kid can be sitting safely at home, doing absolutely nothing wrong, and still get hit with the emotional equivalent of a hallway whisper: *Everyone has this. Everyone is going there. Everyone is becoming that.*

That's the modern twist. Social pressure used to require actual proximity-cafeterias, locker rooms, birthday parties. Now it arrives on a glowing rectangle, customized to your child's insecurities with the patience of a machine that never sleeps.

If you've ever watched your child's mood shift after ten minutes of scrolling-lighter, then tighter, then suddenly angry or down-you've seen the mechanism at work. It's not just *too much screen time*. It's the constant invitation to compare, to upgrade, to keep up. And comparison is expensive.

Money-wise, this is where the spending plan and the 24-hour pause earn their keep. But the deeper skill is psychological: helping your child recognize when a purchase urge isn't really about the product. It's about belonging, status, fear of missing out, or the quiet panic of feeling behind.

The new ad: it doesn't look like an ad

Kids can spot a commercial on TV. They cannot reliably spot a paid recommendation delivered with a smile by someone who feels like a friend.

Influencer marketing works because it borrows trust. The message isn't "Buy this." The message is "This is part of *my life*." The product is blended into a morning routine, a "get ready with me," a bedroom tour, a gaming livestream, a *haul* with a pile of packages like trophies.

Even when there's a tiny disclosure—#ad, "partner," "thanks to ___ for sponsoring"—kids and teens often process the vibe more than the fine print. The vibe says: *This product equals coolness. This product equals friends. This product equals becoming a version of yourself people admire.*

That's not a moral failure on their part. It's a very human response to social cues. Your job isn't to make your kid immune to persuasion. Your job is to make persuasion *visible*.

Because once something is visible, it's easier to resist. A magic trick stops being magic the moment you see the hidden string.

The emotional hook: what the product is really selling

A useful family practice is to talk about ads like you're analyzing a story. Every good story has a problem and a promise.

When your child wants something they saw online, ask one question that sounds curious, not suspicious:

> *"What is this trying to make you feel?"*

At first, your child might say, "I don't know." That's fine. Offer a menu of feelings, like a detective laying out clues:

- *Belonging*: "If you have this, you're in."
- *Status*: "If you have this, people will notice."
- *Security*: "If you don't have this, something will go wrong."
- *Attractiveness*: "If you have this, you'll look better."
- *Escape*: "If you have this, you won't feel bored/sad/stressed."

Most kid purchases driven by social media fit one of those. The product is often secondary. The feeling is the payload.

This approach does something subtle and powerful: it separates your child from the urge. Instead of "I am a person who needs this," it becomes "I'm feeling pulled toward this because it promises something." That little bit of distance is the beginning of self-control.

FOMO is a clock someone else controls

FOMO—fear of missing out—sounds silly until you're the one missing out.

For kids and teens, missing out can feel like social death. Adults might say, "You'll forget about it in a week." Kids know, correctly, that *this week* is their whole universe.

Marketers weaponize that sense of time with countdowns, limited drops, "last chance," "going fast," "only a few left." Platforms weaponize it with disappearing stories and streaks. The message is: *If you don't act now, you lose.*

One of the most helpful things you can teach your child is that urgency is not always real. Sometimes urgency is just a sales tactic dressed as an emergency.

Here's a line that works surprisingly well with kids who are old enough to roll their eyes:

> *"If someone is rushing you, it's usually because rushing helps them."*

Then connect it to the practical tools they already have: the spending plan and the pause. "Yes, if" becomes "Yes, if you still want it tomorrow and it fits the plan." The pause doesn't make your child miss out on everything. It makes them miss out on the things they were going to regret anyway.

Case study: the Stanley cup (or the *whatever* everyone has this year)

Every few years, a specific item becomes a social password. It doesn't matter which one: a particular water bottle, a brand of sneakers, a hoodie with a certain logo, a bag charm, earbuds, a skincare product, a gaming accessory. The object changes, the mechanism stays the same.

Let's call it the *Cup*.

Your middle-schooler, Ava, comes home and says, "I need the Cup." Not "I want." *Need*. Because everyone has it. Because someone at lunch commented on the one she already owns. Because a girl she follows posted an unboxing with a squeal, and the comments were a chorus of heart eyes.

If you say, "That's ridiculous," Ava hears, "You don't understand my life." If you say yes immediately, you may be paying for a lesson that says, "Pressure works."

A smarter move is to treat this as a two-part problem: social belonging and money.

First, validate the social reality without validating the purchase as the only solution:

> *"It makes sense you'd want what your friends are into. Feeling left out feels awful."*

Then move to curiosity:

> *"What do you think having the Cup would change?"*

Ava might say, "People would stop making comments." Or "I'd look normal." Or "I could sit with them." That answer is gold because now you're talking about the real goal: belonging, peace, confidence.

Then you bring in the spending plan:

> *"If you decide you still want it after 24 hours, we can figure out which bucket it comes from. What would you trade?"*

Now Ava is making an adult decision: trading one desire for another. If she chooses to buy it, you haven't "given in." You've made it conscious and bounded. If she chooses not to, she gets to feel a different kind of pride: not the pride of owning the object, but the pride of not being controlled.

What often happens, interestingly, is that the pause changes the story. The first day, the Cup feels like oxygen. The next day, it might feel like "fine, but maybe not $45 fine." Or it might still feel important, and Ava chooses it—and at least she knows what she's paying for.

That's savvy.

The "highlight reel" problem: kids compare their behind-the-scenes to someone else's best frame

Social media isn't fake in the simple sense. It's curated. It's the best angle, the best lighting, the best moment, with the awkward parts edited out.

Kids know this intellectually. They still feel it emotionally.

If your child watches a creator who seems to have endless new clothes, a perfect room, constant trips, and a life that never includes acne, boredom, or loneliness, the comparison can become a kind of itch: *Something is wrong with my life. Maybe the thing I'm missing is stuff.*

Here's the quiet truth worth saying out loud at home: a lot of online "normal" is sponsored, gifted, filtered, staged, or financed by debt. You don't have to accuse anyone. You can simply widen the picture.

Try a conversation like this:

> *"When you see that, do you think it's their everyday life-or their best day?"*

Or, with teens who appreciate bluntness:

> *"There's a reason they call it content. It's not a diary. It's a product."*

That reframe protects your child from confusing someone's performance with your family's reality.

Disclosures: a tiny label isn't a force field

Many sponsored posts include disclosures. That's good-but it's not enough.

A quick, practical way to explain disclosures to kids:

> *"A disclosure is the creator saying, 'I'm being paid or getting free stuff.' It's a clue. It doesn't automatically*

> *mean the product is bad. It means we should be extra careful."*

Then teach them what "extra careful" looks like:

- Don't buy on the first exposure.
- Look for non-sponsored reviews, especially the negative ones.
- Ask, "Would I want this if no one else knew I owned it?"

That last question is especially powerful with status-driven items. It doesn't shame your child; it invites honesty. Sometimes the truthful answer is, "No." Sometimes it's, "Yes, I just like it." Either way, your child is making a decision from the inside out.

Build "ad immunity" with a family game: Decode the pitch

If you want to make this skill stick, make it playful.

Once a week, bring one ad to the dinner table-something from your kid's feed, something from your own, a billboard, a YouTube pre-roll. Then do a quick decode:

- What's the hook? (status, belonging, fear, convenience, humor)
- What's the promise? ("You will be more ___.")
- What's the proof? (real evidence, testimonials, vague vibes)
- What's missing? (price over time, downsides, limitations)

Your child gets to be the expert. You're not delivering a warning; you're building a skill. Kids love skills that make them feel smarter than the system.

Also: do it with your own ads sometimes. Let your kid see you get tempted. "This kitchen gadget makes me feel like a competent adult. I am not, in fact, a competent adult. I am a person being marketed to." A little humor lowers defensiveness and makes skepticism feel normal.

Social pressure: the money conversation that isn't about money

Not all pressure comes from ads. A lot comes from friends.

Sometimes it's direct: "Why don't you have ___?" Sometimes it's indirect: group plans that cost money, inside jokes tied to a product, teasing that lands like a jab.

You can't buy your child out of social discomfort. But you can give them scripts. Scripts are not cheesy; scripts are tools for anxious moments when a kid's brain goes blank.

A few that work:

- *The delay script:* "I'm saving for something else. Maybe later."
- *The boundary script:* "My parents are strict about spending. I'm not doing that."
- *The humor script:* "My wallet says no."
- *The redirect:* "I'm in-just not buying the extra stuff."

These phrases let your kid keep dignity without needing to explain your family finances. They also teach a deeper lesson: you can belong without matching every purchase.

If your child is anxious about saying no, practice out loud once at home, like rehearsal. Athletes practice simple moves until they're automatic; kids can practice social moves the same way.

Curate the environment: you don't have to fight every battle

One of the most practical parenting moves is to admit that no one can withstand nonstop persuasion forever. Adults can't. Kids definitely can't.

So reduce the input.

That might look like:

- fewer shopping apps
- fewer "haul" accounts
- turning off certain notifications
- creating phone-free times where the brain can settle

This isn't about being anti-technology. It's about being pro-mental-space. A kid with breathing room makes better decisions.

A surprisingly effective family question is: *"Does this account make you feel better or worse after you watch it?"* If the answer is consistently "worse," you're not banning joy; you're removing poison disguised as entertainment.

A parent's role: calm translator, not moral judge

When a child wants something because of social pressure, the easy parental move is to judge the desire: "That's shallow." But shallow isn't the real story. The real story is often fear-fear of being left out, laughed at, unseen.

When you treat the desire like an emotion worth understanding, your child learns something rare: feelings can be handled without being obeyed.

You can say, honestly:

> *"I can't buy everything the internet tells you to want. But I can help you think it through so you don't feel pushed around."*

That sentence turns you into an ally against the pressure, not the person who *is* the pressure.

Actionable takeaway: the "3-Question Ad Filter" for any purchase urge

The next time your child wants something they saw online, use this three-question filter before money enters the conversation:

1. *What feeling is this selling me?* (belonging, status, security, attractiveness, escape)
2. *Would I still want it if no one else saw it?*
3. *Can I wait 24 hours and check my spending plan before deciding?*

If your child can answer those questions, they're no longer a passive target. They're a person making choices on purpose—under pressure, in the real world, with their dignity intact.

4.5 Intentional Giving and Generosity

Most kids' first experience with giving is accidental. A coin drops into a donation jar because it's there. A school fundraiser form shows up in a backpack like a surprise quiz. A relative says, "Put a dollar in the church basket," and your child complies the way they comply with passing the salt.

That kind of giving isn't bad. It's just thin. It teaches that generosity is something you do when an adult prompts you, with whatever you happen to have in your pocket.

Intentional giving is different. It teaches that money can be more than consumption—that it can be a tool for belonging, empathy, and influence. It also teaches a subtle kind of confidence: *I can do something about the world I live in.*

And because kids are kids, it works best when it feels concrete. Not like a guilt trip. Not like a vague lecture about "helping people." Concrete: a person, a place, a problem they can picture, and a plan they can follow.

Generosity is a habit, not a mood

A lot of adults give when they feel inspired: after a moving video, during the holidays, when a friend runs a marathon for charity. That's meaningful, but it's sporadic. Kids learn from what's repeated.

If you want generosity to stick, build it into the same structure you use for spending and saving: a small, predictable category. That's the quiet genius of having a Giving bucket in a spending plan. It tells your child, week after week: *Giving is a normal thing we do with money, like buying snacks or saving for goals.*

For a younger child, that might be 50 cents a week. For an older child, it might be 5% of their allowance or earnings. The amount matters far less than the rhythm.

Rhythm turns "nice idea" into identity.

The ``warm glow'' isn't the point---but it helps

There's a reason giving can feel good. When people use money to help others, many report a boost in happiness—a "warm glow" that's real enough to show up in experiments. But it's not automatic, and it's not a vending machine where you insert dollars and receive joy.

Kids are especially sensitive to whether giving feels *chosen* or *forced*. Forced giving often creates resentment. Chosen giving creates pride.

So aim for the lesson that lasts: not "giving will make you happy," but "giving is one way you can live out what you care about." Happiness may show up as a side effect—like a little internal high-five—but meaning is the sturdier reward.

Start with a question most kids can answer

Adults sometimes choose causes the way they choose vegetables: based on what they think they *should* care about. Kids care about what they can imagine.

A better starting point is a question your child can actually answer without faking it:

> *"What do you wish was better in the world?"*

You might get answers like:

- "Animals shouldn"t be hurt."
- "No one should be hungry."
- "Kids shouldn"t get bullied."
- "The park is always dirty."
- "People shouldn"t be cold outside."

Those answers are imperfect and emotionally honest, which is exactly what you want. From there, you can help your child translate a feeling into a cause.

If your child can't name something big, go smaller: "What"s a problem you"ve noticed at school?" or "What"s something that makes you mad-sad?"

The goal isn't to produce a miniature philanthropist with a five-year strategy. The goal is to connect giving to empathy in a way that feels personal.

Teach giving the same way you teach shopping: do a little research

Once your child has a cause, make research part of the giving habit. Not heavy research—kid research.

If you can, pick two or three organizations addressing the same issue and compare them. You're teaching a powerful idea: good intentions still benefit from good information.

Here are kid-friendly questions that keep it practical:

- *What do they actually do?* (Feed people? Rescue animals? Provide tutoring? Build shelters?)
- *Who do they help?* (Local community? A specific group? A region?)
- *How can we tell they"re real?* (Website, contact info, real photos, clear description, reputable partners)
- *Can we see the impact?* (Stories, numbers, updates)

For a teen, you can add one more layer: "Do they share basic financial information?" Many legitimate nonprofits publish annual reports or summaries. The point isn't to make your teen an auditor; it's to teach that transparency is a sign of seriousness.

This echoes the savvy-shopping skill: don't buy the vibe, buy the reality.

A concrete case study: the ``birthday giving" experiment

Consider a real family scenario that's simple enough to copy.

Nina turns 10 and decides she wants a birthday party. She also saw a video about animals being abandoned at shelters, and she can't stop thinking about it. Nina's parents offer a deal: Nina can choose a shelter to support, and the family will match whatever Nina gives from her own money up to $25.

Nina looks at three local shelters online. One has heartbreaking photos but little information. One has clear adoption info, volunteer instructions, and updates. One is farther away and seems to focus on a specific breed.

Nina chooses the shelter with clear updates. Then she decides to do something that turns giving from "money disappears" into "money becomes action": instead of asking friends for gifts, she asks for pet food donations. She writes a short note for the invitation: "If you want to bring something, bring a can of cat food or dog food. If not, just come."

On party day, a small mountain of cans appears. Nina delivers them with her parents, meets a volunteer, and sees the place. It's not glamorous. It smells like disinfectant and wet fur. Nina is both sad and proud.

Two weeks later, Nina says something you can't force: "Can we do that again sometime?"

That's intentional giving. Not because it was huge. Because it was *connected*: cause, research, action, and a real-world feedback loop.

This kind of giving doesn't just teach empathy; it teaches agency. Nina didn't merely feel bad. She did something.

Broaden the definition: money, time, and talent

Some families can give money comfortably. Some families can't. Kids notice, and they sometimes interpret "we can't give money" as "we can't be generous."

That's a painful misunderstanding—so correct it early.

Generosity has at least three currencies:

- *Money*: donations, buying supplies, sponsoring an event
- *Time*: volunteering, helping a neighbor, showing up consistently
- *Talent*: tutoring, making posters, organizing, building, performing

Time and talent are not "second best." In many communities, they are the scarce resource.

If your child is young, "time" might look like helping you sort pantry items for a food drive. If your child is a teen, "talent" might look like designing flyers for a local fundraiser, helping younger kids with homework, walking dogs at a shelter, or joining a cleanup event.

A key lesson here is dignity: you're not "saving" people; you're contributing to a community.

How to budget for giving without turning it into a tax

If giving feels like a forced tithe, kids either comply resentfully or rebel. Either way, they don't learn generosity as a personal value.

Instead, build giving into the plan with two flexible rules:

- *Small and steady beats big and guilt-driven.*
- *Your child gets a real voice in where it goes.*

If your child wants to give more one month because a disaster happened or a friend's family is struggling, that's a great moment to show trade-offs in a compassionate way: "If you give more, what will you give up from Fun or Goals?" This isn't meant to discourage giving. It's meant to honor it as a real decision worthy of planning.

And sometimes, your child will choose not to give in a particular week. That doesn't mean the habit failed. It means the habit is real life, not a performance.

Scam-proofing: teach kids to be generous *and* safe

One of the crueler facts about the world is that kindness attracts scammers. The more emotional the situation, the more likely someone tries to exploit urgency: "Give now," "Time is running out," "We need gift cards," "Send money by wire," "Crypto only."

Kids and teens—especially empathetic ones—are vulnerable to this because they want to help fast. They don't want to ask skeptical questions because skepticism can feel mean.

So give them a framing that protects both their heart and their money:

> *"Checking first doesn"t mean you don"t care. It means you care enough to make sure your help actually helps."*

Then give them a few safety rules simple enough to remember:

- *No giving under pressure.* Urgency is a red flag.
- *No unusual payment methods.* Gift cards, wire transfers, and crypto demands are major warning signs.

- *Verify before you donate.* If a message claims to be from someone you know, confirm through a different channel.
- *Ask an adult before donating online.* Not because kids can't learn, but because it's a good safety boundary.

This protects your child from a particular kind of heartbreak: the feeling of being tricked when they were trying to be good.

Recordkeeping: a surprisingly powerful grown-up habit for kids

Even if your family never itemizes deductions on taxes, donation recordkeeping is still worth teaching. It signals that giving is real enough to track—like any important money decision.

For younger kids, recordkeeping can be as simple as a "giving page" in a notebook:

- Date
- Organization or person helped
- Amount or item
- Why they chose it

For older kids and teens, keep receipts or confirmation emails in one folder. If they donate $250 or more at once as they get older (less common, but possible with jobs or big fundraising), they should know that larger donations often require written acknowledgments to be treated as deductible in certain tax situations. You don't need to turn dinner into tax class. You can simply teach: "Big gifts deserve paperwork."

Paperwork is not the opposite of generosity. It's respect for the fact that money is a tool, and tools are handled carefully.

The quiet character lesson: giving as identity, not performance

Kids live in a world where good deeds can be posted. Sometimes generosity turns into content: "Look what I did." That doesn't automatically make it bad—public giving can inspire others—but it can distort motivation.

A gentle family rule helps: focus on the recipient, not the applause.

You can say:

> *"If we share this, it should be to help the cause, not to collect compliments."*

That's not about shaming pride. Kids deserve to feel proud. It's about teaching a deeper kind of pride: the kind that comes from acting according to your values even when nobody is watching.

Actionable takeaway: the ``Giving Night" ritual (20 minutes, once a month)

Pick one evening a month and make giving a calm, predictable ritual:

1. Your child checks their Giving bucket balance (money, or time they can offer).
2. They choose one cause or person to support this month.

3. Together, you do a quick verification check (is the organization real, is the ask legitimate, is the payment method normal).
4. Your child makes the gift or schedules the volunteering.
5. They write one sentence in a giving log: "I gave because ___."

Over time, those one-sentence reasons become a portrait of your child's values. And values—unlike trends, ads, and social pressure—don't go out of style.

Chapter 5

Modern Banking and Digital Safety

Money used to be heavy in your pocket; now it's just data in the cloud. This shift makes spending easier but safety harder. In this chapter, we build the digital infrastructure your child needs-from opening their first checking account to locking down their passwords-so they can navigate the invisible economy without losing their hard-earned cash to fees or fraudsters.

5.1 Banking 101: Accounts and Interest

A lot of kids have a single, powerful belief about money: if it's in *my* hand (or my app), it's spendable. That belief makes perfect sense when your financial life is a birthday card, a lemonade stand, and maybe a cash envelope in a drawer. Modern money doesn't work like that. Modern money is a system of *buckets* with rules—some buckets are meant to be splashed around, and some are meant to sit still and quietly grow.

If you can teach one simple mental picture early, make it this: a checking account is a *hallway* where money moves. A savings account is a *vault* where money rests. The mistake that creates most early "How did I run out?" moments isn't math—it's mixing the hallway and the vault in your head.

Right around the time your kid gets a debit card, a teen checking account, or their first paycheck, money becomes dangerously convenient. It can vanish with a tap. It can be spent while standing in line. It can feel like a video game currency—numbers changing without the weight of cash in your pocket. That's why the checking-versus-savings distinction matters so much: it puts friction and purpose back into the system.

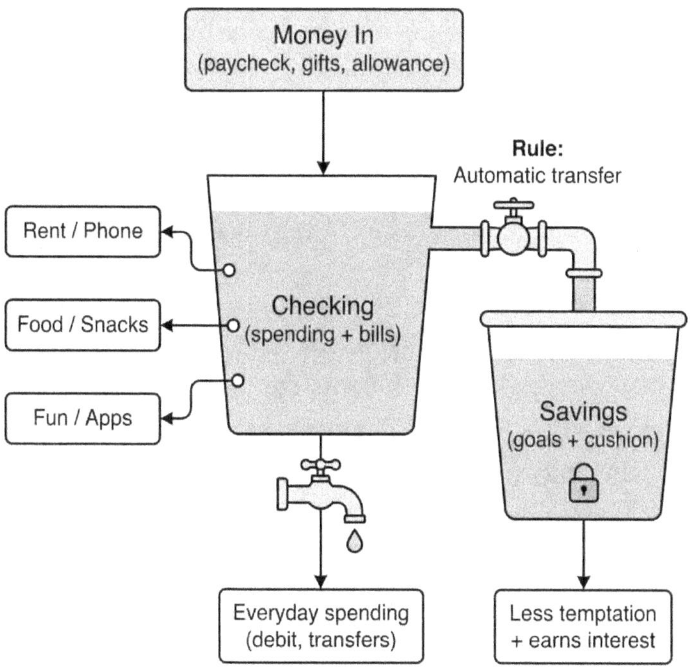

The bucket idea sounds simple, but it changes behavior because it changes the question your kid asks. Instead of "Do I have money?" the question becomes "Which bucket is this money *for*?" That one sentence can prevent a weekend of fun from quietly eating the money that was supposed to cover a field trip, a car insurance payment, or the inevitable *something broke* moment.

Checking: the moving money. Checking accounts are built for transactions: money comes in, money goes out. They're designed to be connected to a debit card, bill pay, direct deposit, and transfers. Checking is the account you expect to touch often, like a kitchen counter. Stuff gets used there.

Savings: the resting money. Savings accounts are built for holding and growing. You can still move money in and out, but the point is different: it's where you park money that has a job tomorrow. A first car fund. A "buy it later" goal. An emergency cushion. Savings is your refrigerator: you don't open it just because you're bored; you open it because you're making an actual meal.

The parent trick is to make these roles feel physical, not abstract. Many kids won't change behavior because you explained a concept. They change behavior because you made a system that makes the right choice easier. When money lands in checking and instantly *looks spendable*, the system is whispering, "Use me." When some money is moved—automatically—into savings, the system whispers, "Protect me."

A concrete example: Jordan is 16 and gets paid every other Friday from a part-time job. The paycheck hits checking: $320 after taxes. Jordan's phone bill ($40) is due in five days. Gas for the week is usually $30. There's also a friend's birthday dinner tonight. In Jordan's head, the banking app shows $320, and the brain translates that into: "I'm good."

Now add the bucket system. Jordan sets an automatic transfer for payday: 10% of each paycheck goes to savings. So $32 moves into savings immediately. Checking shows $288. That number is still enough for everything, but it's a quieter number. It forces a tiny pause: "If I spend $60 at dinner, I'm not spending *money I have*. I'm spending *this week's money*." That pause is where good decisions live.

Notice what didn't happen: nobody gave a lecture about responsibility. The system did the teaching.

That brings us to the other word that shows up in every bank conversation: *interest*. Interest is the rent money pays you for letting someone else hold it. You give the bank your dollars to keep safe and usable, and in return the bank may pay you a little interest. On the flip side, when you borrow money (a car loan, student loan, credit card), you pay interest because you're renting someone else's dollars.

Kids often hear "interest" and assume it's either magic or pointless. The truth is more interesting: interest is powerful *when it has time*, and underwhelming *when it doesn't*. A teen savings account won't turn $50 into $500 by prom, but the habit of understanding interest—how it's calculated and why it's real—sets up adulthood.

Two terms matter here: the *interest rate* and the *APY*. The interest rate is the raw rate the bank pays. APY stands for Annual Percentage Yield, which tries to answer the more honest question: "How much will I actually earn in a year *after compounding*?" Compounding means earning interest not only on your original money, but also on the interest you already earned. Interest-on-interest is the part people get excited about.

Here's the kid-friendly way to explain compounding: it's like planting a garden that drops seeds. The first year you get tomatoes. The second year, you still have your original plants, *plus* new plants from the seeds. That's why time is the superpower. Without time, compounding is a houseplant. With time, it can become a backyard.

Do a simple, honest math moment at the kitchen table. Suppose a teen has $500 in savings and the account earns 4.00% APY (rates change over time, but pick a number that makes the point). Roughly, 4% of $500 is $20 in a year. That's not life-changing, and it shouldn't be sold as life-changing. But now connect it to reality: $20 is a movie ticket and snacks. Or two months of a music subscription. Or a chunk of a new pair of shoes. Interest is small at first, but it's *earned money that didn't require extra hours at work*. That's the seed of the investing mindset: "My money can help me."

The more important lesson is the habit: savings is the place where money can earn. Checking is the place where money can leak.

And checking *does* leak—sometimes in ways that surprise families. The most common leak is ordinary spending, but there are sneakier ones:

- *Minimum balance rules.* Some accounts require you to keep a certain amount in the account to avoid a monthly fee or to earn the advertised rate. The trap is psychological: the bank app shows a big number, but part of it is effectively "do not touch" money if you want to avoid fees.

- *Timing.* Money can leave checking later than you expect. You buy something today, and the transaction might not fully finalize until days later. (That's coming soon in the statement-and-balances conversation.) If your kid spends until the balance is near zero, a delayed charge can push the account negative.
- *Overdrafts.* An overdraft is what happens when a purchase or payment goes through even though there isn't enough money in checking. Some banks cover it (and charge a fee), some decline the transaction, and some move money from savings to checking. Families should decide in advance what they want. For many teens, the safest training wheels are: keep overdraft coverage off for everyday card purchases so transactions get declined rather than turning into fees, and keep a small cushion in checking.

The point isn't to fear banks. The point is to treat accounts like tools with settings. A power drill is useful; you still want the guard on when your kid first uses it.

Another foundational piece to teach early is what it means when money is *insured*. In the U.S., many bank deposits are protected by government-backed insurance (FDIC for banks, NCUA for credit unions) up to certain limits per depositor, per institution, per ownership category. For a kid, the takeaway is simple and calming: a legit insured institution is not the same as cash in a shoebox. If a bank fails, insured deposits are meant to be protected up to those limits. It's the grown-up version of "your money is safer here than under the mattress."

That safety lesson matters because kids live in a world full of money look-alikes. Prepaid cards, payment apps, online "wallets," even game platforms that store balances—many of these feel like "banking," but they don't always come with the same protections, the same clear statements, or the same rules. A checking and savings account at an insured institution is the boring foundation that makes everything else safer.

So how do you turn this into something a kid will actually do?

Start with the bucket rule and make it behavioral, not philosophical:

Rule: Checking is for the next 7–14 days. Savings is for anything beyond that.

That rule forces planning without requiring a spreadsheet. If your kid gets paid weekly, checking holds one to two weeks of spending. If they get paid every other week, same idea. The exact number isn't sacred; the habit is.

Then build the two simplest systems families can run without drama.

- *System 1: Automatic overflow.* On payday (or whenever money comes in), set an automatic transfer from checking to savings. Even $5. Even 1%. The goal is to make saving the default, not the heroic act that happens only when your kid feels unusually mature.

  If your kid resists, try the "small enough to ignore" approach: a transfer so tiny it barely registers. The mind adapts. Once they've lived with it for a month, increasing it feels less like a loss and more like a normal setting.

- *System 2: Two balances in your head.* The bank balance is one number. Your *spendable* balance is another: it's the bank balance minus the money that already has a job (upcoming bills, planned transfers, minimum balance needs, and a cushion). Adults who avoid overdrafts are rarely "better at math." They just keep two numbers in their head, even if the app shows one.

  If you want to make it concrete, give your kid a name for that cushion. Call it the "Oops Fund" inside checking—$50, $100, whatever fits your household. The purpose isn't to create a secret pile to spend later. It's to prevent a small mistake from turning into a fee plus embarrassment plus a frantic parent text.

A real-world case study you can borrow: a parent and teen set up the first checking account with a debit card for school lunches and gas. Everything goes fine for two months. Then the teen buys new headphones online for $79. The purchase "goes through," the app balance drops, and they feel in control. Two days later, a recurring subscription renews for $12.99 (the teen forgot it existed), and a fast-food purchase from the weekend posts late for $14.50. Suddenly the account goes negative by a few dollars, and the bank charges a $35 overdraft fee. The teen didn't "overspend" in the usual sense—they got ambushed by timing and forgetting.

Now replay the same story with the bucket system: payday triggers an automatic $25 transfer to savings, and checking always keeps a $50 "Oops Fund." The late-posting charges still happen. The subscription is still annoying. But the account doesn't go negative, nobody pays a $35 punishment fee, and the problem becomes what it should be: a gentle lesson in tracking, not a crisis.

This is also where you can teach something kids secretly love: *systems beat willpower*. When your kid says, "I'm just bad with money," you can respond, "No, you don't have a system yet." It's empowering. It moves money mistakes out of the "character flaw" category and into the "settings" category.

Once the two-account foundation is set, bring interest back in—this time with a purpose that feels real. Savings grows best when it's tied to a goal with a picture attached. A laptop. A class trip. A car. A "move-out" fund. The moment savings becomes a *story*, the interest becomes a side character: not the hero, but a helpful friend.

A surprisingly motivating approach for teens is to treat interest like a scoreboard that only counts if you don't cheat. Try this family challenge:

Pick a savings goal and give it a time limit. Every month you don't dip into that goal money, you "earn" the bank's interest plus a small family match (like $1–$5). If you withdraw, the match pauses for that month.

This isn't bribery; it's training. It teaches the adult skill of leaving money alone long enough for it to do its job.

Finally, teach the one sentence that prevents a lot of confusion later:

Checking is for transactions. Savings is for goals. Interest is a bonus, not a plan.

That line keeps expectations healthy. It also nudges your kid toward the bigger truth they'll need as they grow: bank savings interest is great for emergency funds and short-term goals, but big long-term growth usually comes from investing, not from a basic savings account. You don't need to teach investing here to plant the seed—you just need to teach the difference between a safe vault and a growth engine.

A practical takeaway you can use tonight: open your banking app with your kid (or use a piece of paper if they don't have an account yet) and label every dollar in checking with a job—*gas, food, phone bill, fun, Oops Fund*. Then pick one small automatic transfer amount from checking to savings that won't cause stress. The goal isn't perfection; it's creating a hallway for spending and a vault for growing—so money stops being a mystery and starts being a tool.

5.2 Tracking the Numbers: Statements and Balances

The first time a teen says, "But my app said I had money," they usually sound betrayed, not careless. And honestly? That feeling is fair. A banking app looks like a scoreboard. Scoreboards are supposed to be final. You don't expect the points to change on Tuesday because the ref remembered a basket from Saturday.

Bank balances don't behave like scoreboards. They behave like a group chat where messages arrive out of order.

If your kid only learns one grown-up banking truth, make it this: the number on the screen is a *snapshot*, not a promise.

Money can be in motion even when it looks still. Transactions have a life cycle. Deposits can be "there" but not available yet. Charges can happen now and settle later. Once you understand the timing, banking stops feeling like a trick and starts feeling like weather: sometimes inconvenient, usually predictable if you check it.

A good place to start is the moment that causes the most confusion: the gap between *pending* and *posted*.

When you tap a card at a coffee shop, the merchant usually asks your bank, "Is this card real, and is there enough money?" Your bank replies quickly, and a hold may show up right away. That's the *pending* stage. It's the banking equivalent of someone putting their hand on the last cookie and saying, "This one's mine." The cookie hasn't left the plate yet, but you should act like it's gone.

A *posted* transaction is the final version. That's when the merchant and bank finish the paperwork and the charge officially settles. Posted transactions are what show up on statements as final, what count as the official record, and what determine your true ending balance for the day.

The dangerous part is the in-between: a purchase can be real enough to reduce what you can safely spend, while still being vague enough to not fully show its effect. That's how people "accidentally" overspend even when they're checking their balance.

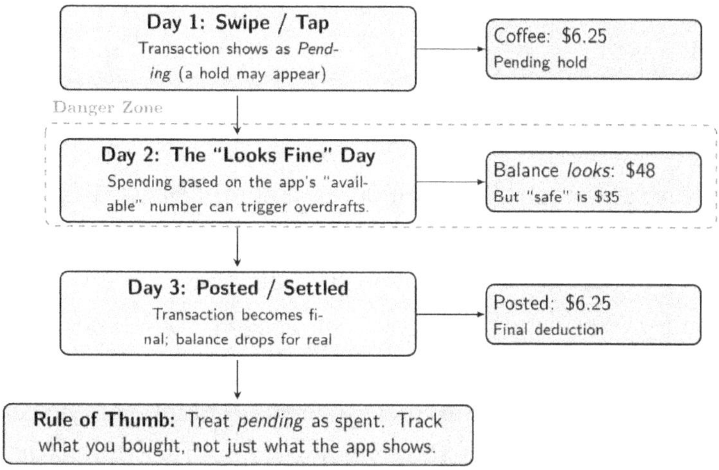

Once kids see this timeline, you can almost hear the gears click: "So the app isn't lying... it's just not finished talking." Exactly.

Now add one more twist that matters in real life: deposits can time-travel too.

A check deposit, a cash deposit, or even a transfer can show up as "received," but your bank may not make all of it available immediately. That's not the bank being dramatic; it's the system managing risk and verification. Banks use rules about *business days* (typically Monday through Friday, excluding federal holidays) and sometimes cutoff times. Deposit at 6:05 p.m. when the cutoff is 6:00 p.m., and it might be treated like tomorrow's deposit. Your kid sees the deposit in the app and thinks, "I'm rich again," then gets a declined transaction at the gas station. It's not personal; it's plumbing.

This is why "available balance" can feel like a trap. People assume it means "safe to spend." Sometimes it does. Sometimes it means "available under current holds, pending items, and deposit rules that may still change." Banking apps are getting better at transparency, but the only fully reliable approach is behavioral: track what you've committed to spend, not just what the app currently says.

A concrete, painfully common case study: Maya, 17, works weekends and gets paid by direct deposit every other Friday. One pay period she also deposits a birthday check from her aunt for $200 using mobile deposit. The app shows the deposit quickly. Maya buys concert tickets that afternoon, assuming the check money is fully there. Later that night, her checking account drops close to zero when several earlier purchases finally post (a streaming subscription, a fast-food charge that posted late, and a ride-share tip that finalized higher than the initial estimate). The concert ticket charge goes through, but a small recurring payment hits overnight and triggers an overdraft fee.

Maya didn't do anything "wild." She did what the app encouraged: she believed the number.

Now replay the same week with one simple family habit: Maya keeps a mini ledger-nothing fancy, just a note on her phone. When she deposits the check, she marks it as "*not spendable yet*" until it's fully available. When she buys anything, she immediately writes it down as spent, even if it's pending. Two minutes a day. No spreadsheets. No lectures. The banking system didn't change; Maya's relationship with it did. The overdraft disappears.

That little ledger is really teaching a deeper adult skill: *reconciling*. Reconciling means comparing two versions of reality and making sure they match-your record of what you did versus the bank's record of what happened.

Banks send you a formal version of their record: the *statement*. Statements feel old-school, like something your grandparents filed in a drawer. But statements are still the official receipt of your financial life. If your kid ever needs to dispute a charge, prove a payment, or catch fraud, the statement is where you go.

Here's the part families miss: statements aren't just paperwork; they're a protection tool.

Many bank accounts that allow electronic transfers provide periodic statements on a regular schedule-often monthly, and at least quarterly in some situations. Those statements list transactions, fees, and balances for a specific period. Even if your kid never touches paper mail, the digital statement (a PDF) is the version that holds up when there's a disagreement.

Statements also create a clock. If an unauthorized electronic transfer shows up, consumer protection rules often depend on how quickly you report it. Waiting "until I have time" is not a neutral choice. The longer you wait, the more complicated it can become—and the more money can slip out the door if someone has access.

That's why the habit you're building isn't "be paranoid." It's "be early."

A practical way to teach this without turning your household into a finance seminar is to make statement review feel like checking a scoreboard *on purpose*, not obsessing. One short ritual works well:

The Weekly Wallet Check.

Pick a day that already has a rhythm-Sunday afternoon, Monday night, whenever-and spend ten minutes with your kid doing three moves:

1. Look at the transaction list (not just the balance) and point out anything unfamiliar.
2. Confirm upcoming obligations: any autopays, subscriptions, or transfers expected in the next week.
3. Compare "what we think we spent" to what the bank shows, so pending items don't become surprise posted items later.

That's it. You're not auditing them. You're training their eyes.

The magic of the Weekly Wallet Check is that it quietly prevents three expensive problems:

First, it catches the "ghost subscription." Kids sign up for a free trial like it's a snack sample at Costco-quick bite, no consequences. Then the trial becomes a charge. Weekly review reveals it while it's still small and recent enough to cancel cleanly.

Second, it catches "double charges" and weird holds. Some merchants (hotels, gas stations, car rentals) place large temporary holds that later adjust. If your kid only watches the balance, it looks like money disappeared. If they watch transactions, they see the story: "hold now, final later." That reduces panic and prevents a knee-jerk transfer from savings.

Third, it catches actual fraud early. Fraud rarely starts with someone draining the account in one dramatic move. It often starts with a small test charge—$1.07, $4.99—something easy to miss. Weekly review is how you spot the test charge before it becomes a $247 "electronics" purchase.

A lot of parents hesitate here because they don't want to be intrusive. That's a reasonable concern-privacy matters, and teens need growing autonomy. The compromise is to treat it like driver's ed. When your kid is learning to drive, you sit in the passenger seat and help them scan mirrors and anticipate problems. You don't do that forever. You do it until they have instincts.

You can also keep it values-based instead of surveillance-based: "I'm not trying to see what you bought. I'm trying to make sure nobody else is buying things *as you*."

To make balances less mysterious, it helps to teach three "bank words" that show up in apps and statements:

Current balance (or ledger balance). Think of this as the running total after posted transactions. It may or may not include pending holds, depending on the bank's display choices.

Available balance. The bank's best attempt at "what you can use right now" after holds and other factors. Useful, but not infallible.

Pending. The waiting room. Real enough to respect, not final enough to relax.

If your kid learns to read those three labels, they stop feeling tricked. They start thinking like a person who handles money professionally—which is a surprisingly empowering identity for a teenager.

Another easy win: alerts. Most banks let you turn on notifications for low balances, large purchases, online transactions, and withdrawals. Alerts are like smoke detectors: you don't install them because you expect a fire; you install them because you want a small problem to stay small.

A good family default is:

- Low-balance alert at a number that protects the "Oops Fund" in checking.
- Purchase alerts for all card transactions (at least while your kid is learning).
- Transfer alerts for any movement out of the account.

If your teen rolls their eyes, remind them: athletes watch game film. Alerts are game film for money.

Now for the part that turns this from "knowledge" into "skill": how to spot an error before it becomes a problem.

Errors tend to fall into a handful of recognizable shapes:

Sometimes it's a merchant error: a duplicate charge, a charge that never should have happened, or a refund that doesn't arrive.

Sometimes it's a timing confusion: a pending item fell off and later reappeared posted; a tip adjusted the final amount; a deposit is partially available.

Sometimes it's your own system failure: you forgot an autopay, mis-timed a transfer, or assumed money was available.

The fix isn't to memorize every banking rule. The fix is to build a habit of asking one calm question when something looks wrong:

Is this unfamiliar, or is this unfinished?

Unfamiliar means you don't recognize the merchant, the amount, or the timing. That's when you act quickly: freeze the card if needed, contact the bank, change passwords if you suspect account access, and document what you see (screenshots help).

Unfinished means you recognize it but it's in progress: pending, a hold, a tip adjustment, a deposit availability issue. That's when you wait a beat, keep your cushion, and watch it move to posted.

Kids like clean answers. Banking is often a "wait and verify" world. You're teaching emotional regulation as much as math: don't panic, don't ignore, don't guess-verify.

One more real-world moment worth teaching: merchants don't always show up on statements under the name your kid expects. "Joe's Tacos" might appear as "JTS Holdings LLC." That sounds sketchy until you realize businesses use legal names and payment processors. The teachable move is to tap the transaction for details, match it to a receipt or memory, and if it still doesn't match, flag it.

This is also where you can set a family rule that prevents a lot of chaos: *Receipts for anything you'd be mad about paying twice.* Not every $3 coffee receipt. But gas, bigger online purchases, event tickets-anything that would cause an argument if it went wrong. Screenshot the order confirmation and keep it for a month. That's not being obsessive; it's being prepared.

All of this adds up to one quiet, adult superpower: the ability to trust money without being fooled by it. Your kid doesn't need to become a human calculator. They need to become a person who reads the story behind the numbers.

A concrete takeaway to use this week: have your kid open their banking app and pick one recent purchase that is still pending (or find one from the past that posted later than expected). Ask them to tell the *three-day story* of that purchase-when it happened, when it showed up, and when it became final. Then set a repeating Weekly Wallet Check on their phone calendar and turn on one low-balance alert that protects a small "Oops Fund." If the app's number ever feels confusing again, that routine turns confusion into a quick investigation instead of an overdraft fee.

5.3 Digital Payments and Apps

A twenty-dollar bill has a built-in speed bump. You have to find it, unfold it, hand it over, and watch it leave. Digital money has no speed bump. It evaporates.

That's the miracle and the danger of modern payments: they remove friction. Friction is what makes you pause long enough to notice what you're doing. When you can pay a friend, split a pizza, subscribe to a game, and tip a delivery driver in the same minute—without ever touching your wallet—your brain stops treating money like a limited resource and starts treating it like a button.

Kids and teens aren't uniquely careless here. Adults fall for the same psychology. The difference is that adults have scar tissue: we've felt the sting of "How did that happen?" and learned not to trust a balance that looks too comfortable. Your job is to give your kid the lesson without requiring the scar.

Start with a helpful reframe: digital payments aren't "spending." They're *permissions*. Every app, tap, and transfer is you giving someone permission to move numbers around in your name. The more places your kid stores permission—the more apps, saved cards, "one-click" buttons—the easier life gets and the more attention it demands.

There are three common ways money moves in a teen's world:

Debit cards. A debit card pulls money directly from checking. If the money is gone, it's gone. Debit is the financial equivalent of handing over cash, except it happens so fast it can feel like nothing happened at all.

P2P apps. Person-to-person apps (think "pay my friend") are built for speed and social life. They're the digital version of passing someone cash at the movies—except now it's searchable, reversible *sometimes*, and complicated when it goes wrong.

Digital wallets. Apple Pay and Google Pay can store cards and make payments even smoother. They're not a separate "account" by default; they're a super-efficient wrapper around a card you already have. They reduce the need to carry plastic, but they also reduce the tiny pause that plastic creates.

The point isn't to ban these tools. It's to teach your kid to treat each one like a different kind of door. Some doors lock behind you. Some doors swing both ways. Some doors look like a door but behave like a trapdoor.

Here's a practical rule that cuts through a lot of confusion:

Debit is for you. P2P is for people you know. Bank transfers are for anything serious.

Debit is a tool for your own spending: groceries, gas, school lunches. It's a clean lane: you buy something, your account decreases.

P2P is a tool for social life: splitting dinner, paying your share of a team hoodie order, reimbursing a parent for concert tickets. It can be perfectly safe when used correctly. The danger is that kids start using P2P the way they use texting—fast, casual, sometimes sloppy—while the consequences are closer to handing a stranger cash.

Bank transfers (or paying by credit card for older teens who can handle it) are what you use when the amount matters, the person is unfamiliar, or you need a clear record. Serious money deserves slower rails.

Why are P2P apps riskier than they feel? Because they blur a line that matters a lot in consumer protection: the difference between *unauthorized* and *authorized-but-a-bad-idea.*

Unauthorized means someone accessed your account without permission—stolen credentials, hacked phone, card number taken. That's closer to theft.

Authorized-but-a-bad-idea means you pressed "send" yourself. You meant to send money, you just sent it to the wrong person, or you were tricked, or the "seller" wasn't real. That's closer to handing cash to someone who lied to you. The system often treats those situations differently. And that difference is the heart of why families need a protocol.

Kids tend to believe all "fraud" is the same. They hear the word and assume the bank simply reverses the charge, like rewinding a video. But a payment you *authorized*—even under pressure—can be much harder to unwind than a payment you didn't authorize at all. Teaching this distinction early is like teaching a teen that not every accident is covered the same way by insurance. It's not fair, but it is real, and knowing it changes behavior.

A real-world case study: Eli, 15, is trying to buy a used graphics card for a gaming PC. He finds a seller in a local group chat who says, "I can meet tomorrow, but I need a deposit to hold it." The seller asks for payment through a P2P app because it's "easier." Eli sends $80. The next day, the seller disappears. No meetup. No item. Eli feels robbed, because he was. But from the payment system's perspective, Eli *told the app to send $80 to that person*. The app did exactly what he asked.

That story is common because it uses three levers scammers love: urgency ("deposit now"), social proof ("I'm in your group chat"), and convenience ("use this app"). It's also a perfect teaching moment because it doesn't require your kid to be gullible. It only requires them to be normal.

The fix is not "don't trust anyone." The fix is to install a speed bump that triggers when money is leaving fast.

Try this family rule, said out loud and practiced:

No money leaves the phone while feelings are hot.

"Feelings are hot" means excited, embarrassed, rushed, scared, or desperate to not disappoint someone. It also means "I really want this deal to be real." When feelings are hot, your kid's brain is not doing careful risk math. It's doing social survival.

A speed bump can be as simple as a 10-minute wait. Or a text to a parent: "I'm about to send $80—sanity check?" Not to ask permission like a little kid, but to borrow an adult brain for thirty seconds. Most scams break when they hit a pause.

Now, the other big psychological trap of digital payments: *micro-spending*. Digital makes small purchases feel smaller. A teen might hesitate before spending $40 in cash, but spend $6.99 five times in a week without noticing. It's not that $6.99 is evil. It's that repeated taps turn into a leak.

If you want to make that lesson land, don't argue about "latte money." Make it visual and personal:

Ask your kid to pull up their last two weeks of transactions and circle anything under $10. Then add it up together. The total is almost always surprising. Not because your kid is reckless, but because digital spending is intentionally designed to be forgettable. The companies want the purchase to feel like a non-event.

This is where the bucket idea from earlier becomes a survival tool rather than a cute metaphor. Checking is the hallway, yes—but digital payments are like doors that open off that hallway. If every door is unlocked and labeled "one tap," the hallway empties.

The fix isn't to lock every door. It's to decide which doors get keys.

A practical way to do that is to create two categories in your house:

Tap-Friendly Spending: things you've already agreed are normal and planned (gas, school lunch, a monthly music subscription that's actually used).

Tap-Resistant Spending: anything that tends to explode when friction is removed (in-app purchases, late-night shopping, game add-ons, "limited-time" online drops).

For Tap-Resistant Spending, make a rule that forces a second step. It can be simple: your kid has to move money into checking from savings first, or they have to wait until morning, or they have to write it down in a note titled "48-hour list." The goal is not punishment. The goal is to make the purchase feel real again.

Subscriptions deserve their own spotlight because they're the sneakiest form of digital spending. A subscription is not a purchase; it's a decision you keep paying for until you stop it. Kids often treat subscriptions like a one-time thing—especially free trials. The adult skill is learning to treat subscriptions like pets: if you adopt one, you're responsible for feeding it every month until it's gone.

A clean family policy: *Every subscription must have an anniversary date.* Once a month, you and your kid look at the list and ask, "Would you buy this again today?" If the answer is no, cancel it immediately. If your kid hates the meeting, make it faster: cancel one thing per month, no debate. It's amazing how quickly "I'll cancel later" turns into a habit of canceling on purpose.

Now, about debit cards specifically: debit feels safe because it uses money you already have. That's true—debit makes it harder to go into long-term debt. But debit also has a sharp edge: if something goes wrong, your money can be tied up while it's investigated. With credit cards (used responsibly), disputes can be handled differently because you're disputing the card issuer's money first, not your rent money.

You don't need to hand a teenager a credit card to teach that concept. You do need to teach the principle: *when money comes straight out of checking, you must protect checking like it's oxygen.* That means:

- Don't keep your whole financial life in one account if you can help it. A checking account is for movement; it's not where you store everything you own.
- Keep a buffer in checking so a mistake doesn't trigger fees or cascade into missed payments.
- Turn on transaction alerts so you learn about problems quickly, not after the fact.

Another subtle trap: "instant" doesn't always mean instant. P2P apps often feel immediate, but moving money from an app back to a bank account can take time (unless you pay for an instant transfer). Teens get caught when they think app money and bank money are the same bucket. They aren't. They're neighboring buckets with a little pipe between them, and sometimes the pipe is slow.

A good habit is to teach "location language." When your kid says, "I have $50," respond with a gentle question: "Where is it—checking, savings, or in the app?" That single question trains their brain to stop treating money as one blob.

Parents also run into a social issue: teens don't want to be the "difficult" friend. If everyone is splitting dinner with a P2P request, your kid doesn't want to say, "Can you do a bank transfer?" That's okay—don't make them weird. Instead, teach them to be *precise*:

Precise beats paranoid.

Precise means: double-check the username before sending. Confirm the last four digits of the phone number. Use the QR code if the app offers one. Add a note describing what it's for ("Team hoodie"). Those aren't dramatic steps. They're adult steps.

And teach one "hard no" that will save them from a ton of scams:

Never send money to a stranger because they sent you money first.

There's a common scam where someone "accidentally" sends money and begs you to send it back. The twist is that the original payment can later be reversed, but your "refund" is real. Your kid thinks they did the right thing and ends up losing money. The safe move is simple: don't play bank for strangers. Tell them to contact the app's support. If it was a real mistake, the platform has a process.

To make these lessons stick, it helps to give your kid a tiny script—something they can actually say without sounding like a hall monitor:

> "I can pay you, but I'm going to double-check the handle because I've heard of people sending it to the wrong person."

Or:

> "I'm not comfortable sending deposits through an app to hold something. I'll pay when we meet."

That's not anxious. That's competent.

Now for the skill that ties all of this together: keeping spending visible.

Digital spending becomes dangerous when it becomes invisible. Your kid can fix that with one habit that takes under a minute a day:

Write it down when you spend it.

Not in a fancy budget app. Not in a spreadsheet. In the Notes app, on a sticky note, anywhere. The act of recording restores friction. It tells the brain, "That was real." It also creates a second record—so when the bank list looks confusing, your kid has their own memory trail.

If your teen is the type who hates writing things down, give them the "photo receipt" method: after a purchase, screenshot the confirmation or take a photo of the receipt and dump it into one album called "Money." Once a week, delete the ones that posted correctly. Keep the weird ones. It's a low-effort way to stay grounded.

A concrete takeaway to use today: sit with your kid for ten minutes and decide which kinds of spending are Tap-Friendly and which are Tap-Resistant in your household. Then, on their phone, turn on purchase alerts for their debit card and set one rule for P2P: *no sending money to a new person until the username/number is verified out loud*. That small speed bump won't make life harder—but it will make the money feel real enough to manage.

5.4 Spotting Fees and Fine Print

Banks have a public personality and a private personality.

The public one is friendly: "Free checking," "No minimum," "Earn interest." The private one lives in the fine print and whispers, "*...as long as you behave exactly the way we expect.*" That's not evil; it's business. Fees are how many financial institutions get paid when customers make predictable mistakes-forgetting, overdrawing, using the wrong ATM, letting an account sit too low, ignoring a monthly statement.

Kids and teens are brand-new to the game, which makes them *perfect* fee targets. Not because they're irresponsible, but because they haven't learned the rules that adults assume. A twelve-year-old doesn't naturally think, "I should read the disclosure about minimum balance requirements to avoid a maintenance fee." A sixteen-year-old doesn't naturally think, "A subscription is a tiny vampire that drinks monthly until I pull it off."

So the goal isn't to turn your kid into a suspicious person. It's to make them fee-literate. Fee literacy is the ability to look at a "free" product and calmly ask: "Free *how*?"

Here's the first big mindset shift to teach: a fee is rarely about the *one* time it happens. It's about what it trains you to do next. Overdraft fees train you to keep more money in checking (or to buy overdraft coverage). Out-of-network ATM fees train you to hunt for the "right" machines. Minimum balance fees train you to store a quiet pile of money you don't touch. Once you see fees as training tools, you stop feeling blindsided and start making deliberate choices.

Begin with a quick, kid-friendly taxonomy: there are *fees for being low*, *fees for being out-of-network*, and *fees for forgetting*.

Fees for being low show up when your account drops under a requirement-often called a monthly maintenance fee, service fee, or minimum balance fee. The bank may waive the fee if you keep a certain daily balance, maintain an average daily balance across the month, or have a direct deposit. Those are three different standards, and families get burned because they assume "I hit the minimum once" is enough. Sometimes it isn't.

Fees for being out-of-network show up when you use the "wrong" ATM. Your bank might charge you. The ATM owner might charge you. You can pay twice for the privilege of getting your own money.

Fees for forgetting are the sneakiest: subscriptions you don't use, paper statement fees, expedited transfer fees, inactivity fees on some products, or "convenience" fees that appear when you're in a hurry and not reading carefully.

If your kid learns to sort fees into those buckets, they'll start spotting them everywhere-like learning to identify poison ivy. It turns a vague warning ("watch out for fees") into a practical skill ("that looks like a low-balance trap").

Now for the fee monster that deserves its own spotlight because it's emotionally brutal: *overdraft*.

An overdraft is what happens when a transaction goes through even though you don't have enough money in your checking account. Banks handle this in different ways. Some decline the transaction. Some cover it and charge a fee. Some automatically pull money from savings or a linked line of credit.

A teen's first overdraft often feels like a prank: they spend $8 they don't have, and suddenly they owe $43. The punishment is wildly out of proportion to the mistake, which is exactly why overdraft fees are such a powerful business model.

Here's the key family insight: overdraft is not a "thing that happens." It's a *setting* you can choose.

Many banks can't charge overdraft fees on ATM withdrawals and one-time debit card purchases unless the customer has opted in to that overdraft service. That detail matters. It means a teen debit card can be configured to *decline* instead of "helpfully" going negative with a fee attached. A declined purchase is embarrassing for thirty seconds. An overdraft fee can linger for weeks, especially if it triggers a chain reaction of more fees.

So what should families choose? There isn't one correct answer. But there is a correct process: decide intentionally.

If your kid is still learning, declining is often the best training wheel. It provides immediate feedback without a $35 tuition bill. The purchase doesn't happen. They learn to check their true spendable balance. They learn to keep a buffer.

If your kid has irregular income or bills that must go through (like a phone bill), you might prefer an overdraft protection link to savings, where the bank transfers money from savings to cover a shortfall. Some banks charge for the transfer; some don't. Either way, it's typically cheaper and less traumatic than an overdraft fee-though you still want the habit of keeping enough in checking so this is a rare event, not a monthly lifestyle.

A line of credit can also be used as overdraft protection, but now you're mixing checking mistakes with borrowing. That can be fine for adults who understand credit, but it's usually not the cleanest learning environment for teens.

A simple case study that shows how fees pile up: Tessa is 18, newly responsible for her own checking account at college. She keeps about $60 in checking "most of the time." One week she buys groceries ($42), then gets gas ($25). The grocery charge posts immediately; the gas station puts a hold that later adjusts. The account goes negative briefly. The bank covers it, charges a $35 overdraft fee, and then a subscription renews for $9.99 while the account is still negative. Another fee. Suddenly Tessa owes $70 in fees over what was basically a normal week.

The real problem wasn't the groceries or gas. The problem was a system with no buffer and a setting that converted small timing issues into expensive penalties.

When families see this once, it becomes obvious why the earlier "Oops Fund" in checking matters. Fees thrive in the absence of slack. Slack is expensive in the short term ("why am I keeping $100 just sitting there?") and cheap in the long term ("oh, that $100 prevented $70 in fees").

Next, tackle the ATM fee monster, because it's the fee that feels the most insulting. You walk up to a machine to access your own cash, and it's like the machine demands a cover charge.

This is where you can teach a very adult skill: *map your money routes*.

If your teen regularly needs cash for a club, tips, lunch money, or a school event, figure out the usual places they withdraw cash and identify in-network ATMs near those places. Many banks and credit unions have apps that show nearby in-network options. If your kid knows where the "safe ATMs" are, they won't pay a $3 fee because they're in a hurry outside a concert venue.

If your teen rarely needs cash, the lesson is even simpler: withdraw cash less often, but in larger amounts (within reason). Five small withdrawals can mean five separate fees.

Then there's the minimum-balance trap, which is the fee version of a hidden tripwire. It often looks like this: "No monthly fee if you maintain a $500 minimum daily balance *or* an average daily balance of $500 *or* have $X in monthly direct deposits."

Parents hear "$500" and think, "Okay, we'll keep $500 in there." Teens hear "no fee" and stop listening. The devil is in the definition. "Minimum daily balance" means every single day. "Average daily balance" means your balance across the month gets averaged-so a few low days can be offset by higher days. Those are different. Kids don't need to master the math, but they do need to learn to look for the *method*.

This is also where the language gets slippery: banks may use terms like "ledger balance" and "available balance," and your kid may assume the bigger number is what matters. Sometimes the bank uses one type of balance to decide whether you earn interest and another to decide whether a fee gets waived. It's not designed to be intuitive. It's designed to be disclosed.

Here's a parent move that works: treat bank account opening like buying a phone plan. You don't need to read everything, but you do need to find the "cost triggers."

If you're opening an account with your kid, ask three questions and write the answers down in a note called "Account Rules":

- "What fee can I get charged monthly, and how do I avoid it?"
- "What actions trigger an overdraft fee, and can we set transactions to decline instead?"
- "Where can I withdraw cash for free (or with reimbursement)?"

Those questions are polite, normal, and revealing. They also teach your kid that financial products are negotiable conversations, not magical boxes.

Now, the sneakiest fee category for modern families isn't even a bank fee. It's the *subscription vampire*. Subscriptions are the perfect predator because they are small, automatic, and quiet. They don't show up as a big "purchase" moment. They show up as a slow drip.

Kids also collect subscriptions differently than adults. Adults tend to have a few large ones (cell phone plan, streaming).

Teens can end up with a long tail: game passes, app add-ons, music, video editing tools, "premium" versions of everything.

The practical skill here is not "never subscribe." It's *never subscribe without a cancellation plan.*

A cancellation plan can be as simple as: the moment your kid starts a free trial, they immediately set a calendar reminder for two days before it renews. Or they cancel right away and rely on the trial period to run out naturally (many services still let you use the trial). Either method prevents the classic teen move: "I'll remember." They won't. Adults don't either.

If you want to make this emotionally compelling, show the "vampire math." A $9.99 subscription feels like nothing. Over a year, it's about $120. Over four years of high school, it's nearly $500. That's not a moral problem; it's a trade-off problem. Suddenly you can ask a values-based question instead of a scolding one: "Would you rather have this subscription, or would you rather have $500 toward a car, a trip, or a new laptop?"

That's how you teach opportunity cost-the invisible cost of the thing you chose, which is always the thing you didn't.

Fee literacy also includes knowing which fees are sometimes negotiable. This is where you can give your kid a powerful first lesson in adult conversation: asking for a waiver.

Banks are more likely to waive certain fees-especially if it's the first time, if the account is otherwise in good standing, and if you ask quickly and calmly. The key is to keep it short and mature. Your kid doesn't need to beg. They need to sound like a person who understands what happened and has a plan.

Here's a script you can practice with your teen. Put it on speaker and let them do the talking while you sit nearby:

> "Hi, I'm calling because I was charged an overdraft fee on ___ date. I understand why it happened, and I've changed my settings so my debit card transactions will be declined if I don't have enough money. This was my first overdraft fee. Is there any way you can waive it as a one-time courtesy?"

That script does three things: it's polite, it takes responsibility, and it signals that the bank won't profit from the same mistake again. Banks like customers who learn.

Even if the bank says no, your kid learned something more valuable than the $35: they learned how to advocate for themselves with a financial institution. That skill transfers to insurance, cell phone bills, medical bills-everything adult life throws at you.

One more advanced-but-worth-it concept: interest can have fine print too. Some accounts credit interest monthly. Some require a minimum balance to earn the advertised yield. Some have rules like "interest accrues daily but is paid monthly," and if the account is closed before the interest is credited, you may not receive it (if properly disclosed). The teen takeaway isn't to fear the details; it's to understand that "earns interest" is not a yes/no label. It's a set of conditions.

So how do you teach all of this without turning dinner into a seminar?

Make it a game of spotting traps in plain English.

Pull up an account page or a fee schedule and ask your kid to highlight every sentence that answers one of these:

- "When do they charge me?"
- "What do I have to do to make it free?"
- "What do they charge me for being in a hurry?"

Kids are good at games. This is a game where the prize is keeping their money.

A concrete takeaway you can use this week: open your kid's bank app (or statement PDF) and hunt for fees like you're looking for leaks under a sink. If you find one, don't just pay it-name the trigger and change one setting or habit to prevent it. Turn on a low-balance alert, choose the "decline if not enough funds" option for debit (if available and appropriate), and write down two in-network ATMs near the places your kid actually goes. Fees love forgetfulness; your job is to replace forgetfulness with a few small, repeatable systems.

5.5 Cybersecurity for Families

A decade ago, a teenager could lose a wallet and it would be a bad day. Today, a teenager can lose a phone and it can be a bad month.

That's not because kids are reckless. It's because the phone is no longer just a phone. It's the house key, the debit card, the mailbox, the yearbook, the photo album, and the front desk at the bank. If someone gets into it, they don't just see your kid's life-they can *act* as your kid: reset passwords, approve transfers, impersonate them to friends, and drain accounts in quiet little sips.

The good news is that most "account takeovers" aren't movie-hacker problems. They're ordinary human problems: reused passwords, weak locks, borrowed devices, and one unlucky click when someone is tired. That means the solution isn't genius. It's hygiene—like brushing teeth, but for logins.

The first mindset shift for families is to stop treating cybersecurity as a technical topic and start treating it as *home security*. You don't need to understand how a lock is manufactured to know you want the door locked. You don't need to speak "computer" to make your kid dramatically harder to rob.

A helpful way to explain the basics to a kid is the "three kinds of proof" idea. When a website or app asks, "Are you really you?" it can check:

- something you *know* (like a password),
- something you *have* (like your phone or a security key),
- something you *are* (like a fingerprint or face scan).

When your kid uses just a password, they're using one kind of proof. If the password leaks, gets guessed, or gets phished, the door opens.

Multi-Factor Authentication (MFA) means using two (or more) kinds of proof. It's the grown-up version of "you need both the ticket *and* the wristband." Even if a thief steals one, they still can't get in.

Here's the castle metaphor that makes MFA click: the password is the drawbridge. It gets you onto the property. MFA is the main gate. It keeps you out of the castle even if you crossed the moat.

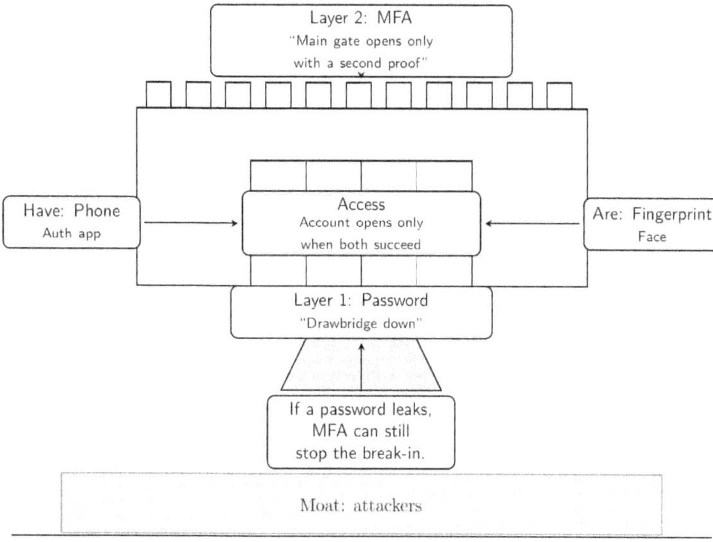

Once your kid understands MFA as a second lock, the rest becomes easier. You're no longer arguing about inconvenience. You're talking about doors.

Now comes the uncomfortable part: passwords are still everywhere, and they are still one of the easiest ways to lose an account. Passwords fail in two big ways.

First, they get reused. A teen makes one "good" password, uses it on a game site, a shopping site, an email, and a bank. One of those places gets breached. The password ends up for sale. Now the thief tries the same email/password pair on other services. This is called *credential stuffing*, and it works depressingly well because people are creatures of habit.

Second, passwords get tricked out of you. This is phishing: a fake login page that looks real enough to fool you when you're tired or rushed. Modern guidance is blunt: passwords are not phishing-resistant. A strong password doesn't help if you type it into a fake page.

So what does help?

Longer, not weirder. Most people think a "strong" password means symbols and chaos. But length is often the bigger win. A passphrase—several ordinary words strung together—can be both memorable and hard to guess. "correct-horse-battery-staple" got famous for a reason: humans can remember words better than random strings, and attackers have a much harder time with length.

Unique, not heroic. The standard you want for your kid isn't "memorize 30 perfect passwords." It's "never reuse." The only realistic way to never reuse is to use a password manager.

A password manager is an app that stores your passwords in an encrypted vault. Your kid remembers one master password (make it long), and the manager generates and fills unique passwords for everything else. This sounds fancy. It's actually less work than the current teen strategy of "I'll remember" plus "Forgot password" every three weeks.

Families often hesitate because putting all passwords in one place feels scary. The reframe: a password manager is like a safe. Yes, everything valuable is inside. That's the point. It's far safer than leaving keys under 30 different doormats.

A good family setup conversation sounds like this:

> "Your job is to have one master password that's long and private. My job is to help you set the manager up and turn on MFA. We're going to make you hard to hack, not perfect."

Now add the device layer, because a phone with a weak lock is like a bank branch with the front door propped open.

If your kid's phone lock is "1234," that's not a lock. That's a polite suggestion.

The most important settings are unglamorous:

A real device passcode. Six digits is better than four. An alphanumeric code is even better. Biometrics (fingerprint/face) are great for convenience, but the underlying passcode is the real key.

Auto-lock quickly. The phone should lock itself fast—one or two minutes. Kids hate this until they lose a phone at school and realize how many hands could have touched it in five minutes.

Find-my-device and remote wipe. If the phone goes missing, you want the ability to locate it and erase it. This is the digital version of canceling cards when a wallet is stolen.

Keep the operating system updated. Updates aren't just new emojis. They patch security holes. A phone that's months behind on updates is like a house with a broken window you keep forgetting to fix.

A case study that shows how these layers interact: Sam is 14 and uses a bank app to check a small account with allowance money. Sam also uses the same password for email and a popular gaming site. The gaming site gets breached. A thief tries the same login on Sam's email, gets in, and uses email access to reset the bank password. If Sam has no MFA on email and no MFA on the bank, the thief can move money or at least cause chaos.

Now replay it with basic hygiene:

Sam uses a password manager, so the gaming password is unique. The breach doesn't help. Even if the thief guesses or steals the email password, the email account has MFA turned on, so the thief can't reset anything without the second factor. Even if the phone is stolen, it has a strong passcode and auto-lock. The thief can't casually open the authentication app. And if the worst happens, the family can remote-wipe the phone.

Same kid. Same world. Completely different outcome.

It's worth saying out loud: the goal of cybersecurity isn't to become unhackable. That's fantasy. The goal is to become *not worth the trouble*. Attackers are usually shopping for easy targets. If your kid has MFA and unique passwords, most opportunistic attacks bounce off and move on.

One more subtle lesson for kids: *where* you put MFA matters.

MFA via a text message code is better than nothing, but it's not the gold standard. Texts can be intercepted in certain scenarios, and phone numbers can sometimes be hijacked through tricks aimed at mobile carriers. Using an authenticator app or a physical security key can be stronger. You don't need to scare your kid with technical details. You can keep it simple: "Texts are okay; an authenticator app is better."

Also teach them that "security questions" are often terrible security. "What's your mother's maiden name?" is not a secret in the age of social media and public records. If a service forces security questions, treat the answers like additional passwords, not like real facts.

Now bring it back to daily life. What do families actually need to do, without turning the home into an IT department?

Create a short "non-negotiables" list and stick to it. Not a poster. Not a lecture. A small set of defaults:

- MFA turned on for email, banking, and payment apps.
- Password manager used for anything that matters.
- Phone locked with a strong passcode and auto-lock.
- Purchase and transfer alerts enabled on financial accounts.

Notice what's missing: "be careful." "Don't click links." Those are good advice, but they're not systems. Systems beat intentions, especially for tired humans.

It also helps to teach kids a phrase that changes how they handle messages:

"Links are not facts."

A text that says "Your account is locked" might be real. It might be fake. The link doesn't prove anything. The safe move is to open the app directly or type the known website yourself. This is a tiny habit that blocks an enormous number of attacks.

If you want to make this feel less like fear and more like empowerment, frame it as competence: "You're learning how to run your own financial life. These are the locks professionals use."

A concrete takeaway you can do tonight: pick one "crown jewel" account-your kid's email, if they have one, because email controls password resets-and turn on MFA. Then install a password manager together and change just *three* passwords to unique, manager-generated ones: email, bank/payment app, and the most-used shopping or gaming account. Finally, set the phone to auto-lock quickly and require a strong passcode. You don't need to win cybersecurity in one evening; you just need to make your kid's digital front door feel like a door, not a curtain.

5.6 The Anti-Scam Playbook

Scams don't start with a question about money. They start with a question about *you*.

Are you busy? Are you kind? Are you embarrassed easily? Do you hate disappointing people? Do you panic when an adult voice says "urgent"? Scammers aren't trying to beat your kid at math. They're trying to steer your kid's feelings like a shopping cart with one bad wheel—pulling hard toward urgency, secrecy, and speed.

That's why the most useful anti-scam lesson isn't a list of "types of scams." Lists go stale. It's a playbook: a few rules that work even when the script changes.

The first rule is the one kids resist until they need it:

Any message can be fake.

Not "any stranger can be fake." Any *message*. The number can be spoofed. The name can be copied. The profile photo can be stolen. The "official" logo can be pasted onto a web page in thirty seconds. A scam in 2026 doesn't have to look shady. It just has to catch you in the moment when you're not expecting to think.

This is also why scams hit teens so hard: their lives are a constant stream of messages. They live in the doorway. Scammers simply walk up and blend in.

So instead of teaching kids to judge whether something "looks real," teach them to judge whether something is *safe to act on*. Safe is about process, not vibes.

A good way to frame social engineering (the fancy term for manipulating people into giving up information or money) is this: scammers don't hack computers; they hack *relationships*. They borrow authority ("I'm the bank"), borrow closeness ("I'm your friend"), or borrow fear ("You're in trouble"). Then they demand speed, because speed is where mistakes happen.

If you want a single sentence that covers most modern scams, it's this:

Urgency is the scammer's favorite seasoning.

They sprinkle it on everything: "Act now." "Don't tell anyone." "This expires." "If you hang up, you'll lose your account." "I'm locked out and I need you." The goal is always the same: bypass the part of your brain that asks, "Wait—does this make sense?"

That leads to the second rule:

No money leaves the family under pressure.

Pressure can be emotional ("I'm crying"), social ("everyone's waiting"), or authoritative ("this is the police"). Your kid doesn't need to identify the scam. They just need to recognize the pressure and hit the brakes.

Here's what "hit the brakes" looks like as an action, not a vibe:

> Pause. Hang up. Close the tab. Take a breath. Verify using a known number or a known app.

Verification means you do *not* reply to the message, click the link, or call the number they provided. You go to the source yourself: the bank's official app, the company's official website typed by you, the number on the back of the card, or a contact saved in your phone *from before the message arrived.*

This is the heart of a Family Verify Protocol. It's not complicated, but it has to be agreed on ahead of time, like a fire drill. You don't invent a fire escape plan while smoke is in the hallway.

A strong Family Verify Protocol has two pieces: permission and procedure.

Permission: Your kid is always allowed to be "rude" to a message. Always allowed to hang up. Always allowed to say, "I'm going to call back." Good kids get trapped because they're trying to be polite. Scammers weaponize politeness.

Procedure: A short checklist your kid can follow even when their stomach is tight.

Here's one that works in real life:

1. *Stop.* Don't click, don't send, don't reply.
2. *Screenshot.* Save the message and any details.

3. *Switch channels.* Contact the person or organization using a number you already trust.
4. *Tell an adult.* If you're under 18, loop in a parent/-guardian *before* money moves.

That's the skeleton. Now flesh it out with the red flags kids can spot quickly.

A scam is usually asking for one of three things: money, access, or secrecy.

Money can be direct ("send $200") or disguised ("pay this fee to unlock your account," "buy something to verify"). If the message asks for money in a weird way, it's a red flag.

Access can be a password, a verification code, a one-time passcode texted to your phone, or remote access to your device. This is the most important detail to drill into kids: *verification codes are not for sharing*. If someone asks for a code that was texted to your phone, they are often trying to log in as you. The code is the key. Handing it over is opening the door.

Secrecy sounds like "don't tell your parents," "don't tell your bank," "this is confidential," "if you talk to anyone, you'll mess it up." Legitimate organizations don't need your secrecy to do legitimate things. Secrecy is for controlling you.

Now add the loudest siren of all: gift cards.

If your kid hears "gift card" in a payment demand—Apple gift cards, Google Play cards, Visa gift cards—that's almost always a scam. Real businesses and government agencies don't require payment in gift cards. Gift cards are the scammer's favorite because they're hard to reverse and easy to move. If you want a single "hard no" rule your kid can remember forever, make it:

Gift cards = stop sign.

A concrete case study shows how these red flags combine.

Nina is 16. She gets a text that appears to be from her "bank" saying there's suspicious activity and her account will be frozen unless she verifies immediately. The text includes a link. Nina clicks. The page looks real: logo, colors, even a "fraud department" number. She types her username and password. Then the page asks for a code that was just texted to her phone. Nina enters the code. The page says "verification complete." She exhales.

Behind the scenes, Nina just handed a thief everything needed to log in for real. The thief uses the code to bypass the bank's MFA and gets access. Within minutes, money is moved out—possibly through transfers that are hard to reverse. Nina checks her account and panics. The scam did its job: it created urgency, pushed her into a link, and tricked her into giving up a code.

The fix to Nina's story is not "be smarter." Nina *was* smart. The fix is a protocol: "If the bank texts you, don't use the link. Open the bank app yourself." And: "Never share a verification code with anyone, ever." Those two habits block the whole chain.

A second case study, closer to teen social life:

Marcus, 15, gets a DM from a friend: "Hey, I'm locked out of my account. Can you send me $40 on a payment app? I'll pay you back tomorrow." Marcus wants to help. He sends it. Then he gets another message: "Actually I need $80 for the locksmith." Marcus hesitates. The friend says, "Please don't tell anyone, it's embarrassing." Marcus sends it. Later he finds out his friend's account was hacked.

Notice how the scam exploited kindness and embarrassment. Teens are allergic to embarrassment. A scammer knows that.

The protocol solution is simple and socially workable: *verify with a second channel*. If a friend messages for money, call them. If you can't reach them, don't send. If it's truly urgent, they can find another solution. Urgency is not proof.

Now bring in the financial mechanics that make early action so valuable. When something suspicious happens—especially unauthorized electronic transfers—time matters. Protections can be stronger when you report quickly. Waiting can turn a contained problem into a cascading one: the thief keeps testing charges, keeps transferring, keeps trying different pathways. Fast reporting is not just emotionally comforting; it can reduce real losses.

That's where the "Weekly Wallet Check" habit becomes more than a budgeting tool. It's an anti-scam tool. Kids who review transactions regularly spot the "test charge" early. Kids who don't can miss the first clue until the problem is expensive.

You can also teach a tiny but powerful behavior: *name the emotion out loud*. This sounds almost too simple, but it works because it breaks the spell.

If your kid says, "I'm freaking out," you can respond, "Good catch. Scammers love 'freaking out.' Let's slow down." If they say, "I feel dumb," you can say, "Scammers love 'dumb' because it makes people hide it. You're not in trouble for telling me."

That last line might be the most important anti-scam tool in a family: a kid who is scared of being punished will hide mistakes. A kid who expects help will report quickly. Speed is protection.

To make the protocol stick, give your kid ready-made phrases they can use without sounding dramatic. Practice them once like you'd practice ordering food in a new language.

> "I can't do this right now. I'm going to verify and call back."

> "I don't click links for banking. I'm opening the app."

> "I don't send money from messages. Call me."

Those phrases are shields. They buy time. Time is the enemy of scams.

Also teach your kid what *not* to do in a panic, because panic creates a second wave of damage:

Don't move money around randomly "to be safe" without understanding what's happening. Scammers sometimes trigger fear specifically to get people to transfer money into a "safe account" that isn't safe at all.

Don't download remote-access software because a "support agent" tells you to. If someone can control your screen, they can often control your money.

Don't pay a fee to "unlock" something. Real organizations have processes, not ransom demands.

Then give them a calm emergency plan that fits on one sticky note:

- If you think an account is compromised: change the password (using the official app/site), turn on or reset MFA, and contact the institution using a known number.
- If money moved that you didn't authorize: report it immediately and document what you saw.
- If you sent money and now suspect a scam: report it to the app and to your bank right away anyway. Even if it can't be reversed, reporting creates a record.

Make "document what you saw" concrete: screenshots, dates, amounts, usernames, phone numbers. Scams thrive on confusion. Documentation restores clarity.

Finally, bring it home with a family agreement that feels supportive rather than controlling:

No one in this family gets in trouble for reporting a possible scam quickly.

That doesn't mean there are no consequences for choices. It means the first priority is containment, not blame. Blame makes kids hide. Hidden problems get expensive.

A concrete takeaway you can do this weekend: write a Family Verify Protocol in one paragraph and save it as a pinned note on your kid's phone. Include one hard rule (gift cards are always a stop sign), one verification rule (use a known number or app, never the link provided), and one emotion rule (if it feels urgent or secret, pause and tell an adult). Then do a two-minute role-play where you text your kid a fake "bank alert" and let them practice the correct response: close it, open the real app, and ask for help before doing anything else. That rehearsal is a small investment that can save real money—and a lot of shame—later.

Chapter 6

Understanding Credit and Debt

Debt is like a chainsaw: a powerful tool that can build a house, or a dangerous hazard that can cause serious injury if handled without respect. This chapter moves beyond the simplistic 'debt is bad' mantra to teach your teen the actual physics of borrowing. We will cover the compounding mechanics of interest, the rules of the credit score game (the 'Adulting GPA'), and how to spot the predatory lenders waiting for them to turn eighteen. The goal is to raise an adult who controls their debt, rather than one controlled by it.

6.1 The True Cost of Borrowing

On a random Tuesday, your teenager might do something completely normal: open a shopping app, see a pair of sneakers, and think, *I can handle twenty bucks a month.* That sentence is where debt starts to get dangerous—not because the kid is reckless, but because the human brain is wired to think in *monthly bites*, not *total meals*.

Borrowing is a machine that turns *time* into *money paid*. The longer the machine runs, the more it produces—and what it produces isn't value. It produces interest.

Here's the trick: most debt doesn't feel like spending. Spending feels like something leaving your hand. Debt feels like nothing happening today. The pain is delayed, sliced into small charges that arrive later, often when you're busy, tired, or not paying close attention. Interest thrives in that fog.

APR: the sticker price on borrowed money (with fine print)

APR stands for *Annual Percentage Rate*. In plain language: it's the price tag on borrowing, expressed as a yearly rate. If a card says 20% APR, your balance is being charged roughly 20% per year *if you carry it.*

Two important clarifications parents can translate for teens:

- *APR is a rate, not a dollar amount.* The dollars depend on how big the balance is and how long you keep it.
- *APR is not always applied once per year.* Credit cards often compound daily or monthly (meaning interest is calculated frequently), so the timing of your payoff matters.

If you want a teen-friendly way to say it: *APR is the rent you pay for using someone else's money. The longer you stay, the higher the rent bill.*

Compound interest: the math that makes small mistakes loud

Simple interest is easy: borrow $1,000 at 20% and you pay about $200 per year in interest. But many real-world debts don't behave that politely.

Compound interest means interest can start charging interest. On a credit card, this happens because interest is added to your balance, and then the next round of interest is calculated on the new, slightly bigger balance. That's why the debt doesn't just sit there. It grows. Quietly, predictably, and without your permission.

This is also why debt and saving are mirror images. When you invest, time is your employee. When you borrow, time is your landlord.

The ``Rule of 72" (in reverse): how fast debt can double

You may have heard the Rule of 72 as a quick way to estimate how long an investment takes to double: divide 72 by the annual rate.

Flip it around for debt. At 20% interest:

$$\text{Doubling time} \approx \frac{72}{20} = 3.6 \text{ years}$$

That number is a jolt for most people. It means a balance you leave alone doesn't just hang around like an unpaid tab. It has the potential to become two tabs in under four years.

Even if the balance doesn't literally double (because most people pay something), the Rule of 72 teaches the right instinct: high-interest debt is a *fast-growing problem*. It doesn't need a dramatic mistake to become dramatic. It just needs time.

Minimum payments: designed to keep you paying, not to help you finish

Here's the part that feels unfair until you understand it: the minimum payment on a credit card is not engineered to get you out of debt quickly. It's engineered to keep the account "current"—meaning you're not late—while stretching the payoff timeline.

Minimum payments are typically calculated as a small percentage of the balance (often around 1%–3%), plus interest and fees. That structure creates a treadmill:

- early on, a big chunk of your payment goes to interest,
- the balance shrinks slowly,
- because the balance shrinks slowly, interest keeps showing up,
- because interest keeps showing up, you keep paying.

If your teen hears "minimum payment" and translates it as "the recommended payment," the credit card has already won a psychological victory. A better translation is: *minimum payment means 'the smallest amount they'll accept to keep charging you.'*

Now for the part that lands: a purchase doesn't cost what the price tag says. It costs what you *ultimately pay*.

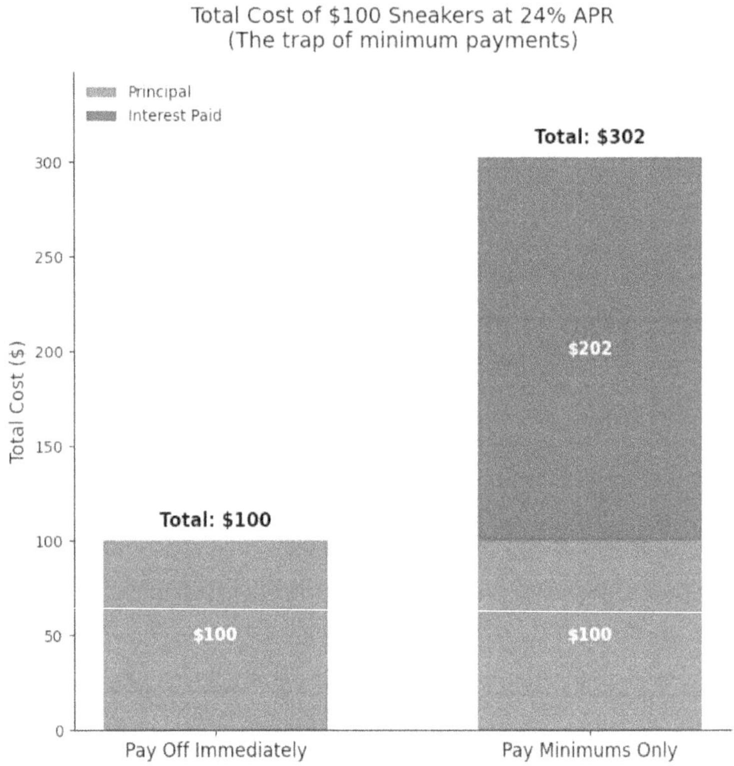

A concrete case study: the sneakers that quietly became $300

Picture a 16-year-old—we'll call him Mateo—who wants a $100 pair of sneakers. Mateo isn't irresponsible. He's just thinking like a normal person: *I'll handle it later.*

Here's how "later" turns into a weirdly expensive shoe closet.

Mateo puts $100 on a card with a high APR (say 24%). He makes a couple of small payments, then gets distracted: a busy month at school, a sports trip, a birthday gift for a friend. He misses a due date once and gets hit with a late fee. Next month, he pays the minimum. The balance is still there, so interest keeps getting charged.

What's the lesson? Not "never buy sneakers." The lesson is that *debt has its own gravity*. Once you're on the treadmill, you have to actively walk off.

A parent doesn't have to scare a teen with a lecture. Just turn it into a number story:

> "You didn't buy $100 sneakers. You bought $100 sneakers *plus* the cost of time. If you pay this off fast, time is cheap. If you don't, time is wildly expensive."

One of the most powerful parenting moves is to make the hidden cost visible *before* the purchase. Teens don't lack intelligence; they often lack a mental model for what interest does when it's left alone.

Why ``monthly payment thinking" is a trap

A lot of bad borrowing decisions don't start with someone saying, "I want to get into debt." They start with someone shopping for a monthly payment.

Car dealers know this. Furniture stores know this. Buy-now-pay-later apps know this. Even adults who are otherwise careful get hypnotized by a small number that repeats.

If you want to inoculate your family against this, teach one sentence that snaps the brain out of monthly-payment mode:

> "What's the *total* I'll pay when all is done?"

That question forces a different kind of math. It shifts the focus from affordability *today* to cost *over time*. It also teaches a deeper life skill: delayed consequences still count.

A quick parent script: translating APR into plain English

When your teen sees an APR, they may not have any emotional response to it. "20%" looks like a grade on a test. Here's a translation that sticks:

> "An APR is how much extra money this purchase demands from Future You. A high APR is a bully: it charges you more the longer you wait."

Then do a short, concrete comparison:

- Pay it off in a year? Interest might be annoying but manageable.
- Stretch it for years? Interest can become the largest part of the bill.

You're not teaching advanced finance. You're teaching a gut-level reflex: *time is expensive when you borrow at high rates.*

The sneaky multiplier: fees and ``one-time" charges

Interest gets most of the attention because it's easy to describe, but fees often provide the shove that turns a small wobble into a fall:

- late fees,
- returned payment fees,

- annual fees,
- cash advance fees (and usually higher interest immediately),
- deferred-interest promotions (where interest can appear retroactively if you miss the payoff deadline).

The teaching moment here is simple: *debt isn't only about the rate*. It's about the *rules*. A card can be "20% APR" and still be far more expensive in real life if the borrower is regularly late, disorganized, or paying inconsistently.

This is why financial capability is partly math and partly habits. A teen who can calculate APR but can't reliably pay on time will still leak money.

A practical way to teach it: turn borrowing into a ``total cost receipt"

Next time your teen wants to buy something with a payment plan, do this together on a phone calculator. It takes two minutes and changes the feel of the decision.

Step 1: Multiply payment by months.
If it's "$30/month for 24 months," the total is $720.

Step 2: Subtract the price.
If the item costs $550, then $720 - $550 = $170. That's the cost of time, before you even talk about APR.

Step 3: Ask what else $170 could do.
Not as a guilt trip—as a trade-off lens. $170 is a month of groceries for some families. It's sports fees. It's a bus pass. It's a

chunk of a laptop. It's a weekend trip. Interest isn't just "extra." It's money that could have built something else.

If you want to go one step further, you can name the trade-off out loud:

> "Borrowing means paying for this item *and* paying the interest item you didn't buy."

That's opportunity cost in plain clothes.

How to make the lesson land without fear

Some kids hear anything about debt and immediately conclude one of two unhelpful things:

- "Debt is evil, I should never use credit," or
- "Debt is normal, everyone has it, who cares."

A healthier target is nuance: *borrowing is a tool that charges rent*. Tools are useful. Tools can also hurt you if you use them the wrong way or at the wrong time.

If your teen is old enough for a phone, they're old enough for the core rule:

> "If you borrow for something that disappears fast, you'll be paying after the fun is gone."

That sentence alone filters out a lot of bad debt: clothes, food delivery, concert tickets, impulse shopping. The pleasure ends quickly; the payments linger.

Actionable takeaway: a one-sentence family rule (and a two-minute check)

If you want a simple household policy that prevents most high-interest trouble, try this:

> *If you can't pay it off within two paychecks, don't borrow for it.*

Then add a quick "total cost check" any time borrowing is on the table:

- What's the total paid? (payment × months)
- How much is interest/fees? (total paid – price)
- Is the item still worth it at the *real* price?

That's the habit that turns credit from a trap into a tool: not bigger willpower, not perfect math—just a reflex to ask what time will do to the bill.

6.2 Credit Cards: Users Manual

A credit card is a weird object when you really look at it. It's a rectangle of plastic that can buy dinner, airline tickets, a last-minute textbook, a birthday gift you forgot, or a streaming subscription you swear you canceled. It works in seconds. It rarely asks you to pause. And if you use it the wrong way, it quietly turns into one of the most expensive loans most families will ever touch.

The goal isn't to make your teen afraid of credit cards. The goal is to make them *bilingual*: able to speak both languages a credit card uses. One language is *convenience* (tap, done). The other is *rules* (billing cycles, statement balances, due dates, fees). Most people only learn the rules after they've paid tuition in the form of interest. Your kid can learn them before.

Two types of cardholders: transactors and revolvers

Credit card companies don't say it this way in their ads, but there are two basic kinds of users.

A *transactor* uses the card like a payment tool: swipes during the month, then pays the statement balance in full by the due date. The bank earns money from interchange fees merchants pay, and the user (usually) pays no interest on purchases.

A *revolver* carries a balance from month to month. That balance 'revolves' forward, and interest starts piling up. The bank earns serious money here.

This isn't a moral label. It's a mechanical one. And it's a powerful family conversation starter because it keeps the focus on behavior, not identity. Your teen doesn't have to be "good with money" as a personality trait. They just need to build the transactor habit: *charge what you can pay off, then pay it off.*

The part most people miss: your credit card has two different ``balances"

Ask a room of adults what they owe on their credit card, and many will open an app and read the big number on the home screen: the *current balance*. It feels like the truth because it's right there.

But for avoiding interest (and for clean, stress-free card use), the number that matters most is the *statement balance*—the total at the end of the billing cycle. That's the amount you can pay by the due date and typically owe *no interest on purchases* (assuming you have a grace period and you haven't been carrying a balance).

So your teen needs a simple mental model:

- *Current balance* = what you've charged so far.
- *Statement balance* = what you owed at the snapshot moment when the statement closed.

A credit card bill is like a photo, not a live video. The statement is the photo. You pay the photo.

Statement date vs. due date: the ``two-door" system

Credit cards run on a rhythm that isn't intuitive at first because there are two key dates.

Statement date (closing date). This is when the billing cycle ends and your statement balance is created. Think of it as the cashier printing the receipt.

Due date. This is the deadline to pay at least the minimum payment (and ideally the full statement balance). Think of it as the "no drama" date. Miss it and you can get hit with fees and damage your credit history.

A useful family script is: "The statement is when they add it up. The due date is when you make it right."

Here's the subtle point that saves people money: if your teen pays the statement balance in full by the due date, they can often use the card for purchases without paying interest. That interest-free window is called the *grace period*. It's not magic. It's a reward for staying in transactor mode.

Autopay: the seatbelt, not the steering wheel

If you want one habit that prevents the most damage, it's this: set up autopay.

Autopay doesn't mean "set it and forget it forever." It means you remove the worst failure mode: forgetting a due date. Late payments are expensive in two ways. There's the obvious pain (a late fee). And there's the long shadow (a late mark can hurt your credit for years).

The key is choosing the right autopay setting. Most issuers offer:

- *Minimum payment autopay* (protects against late fees, but can keep you in debt),
- *Statement balance autopay* (the transactor setting, if the checking account can support it),
- *Fixed amount autopay* (useful if you're paying down a balance intentionally).

For a teen in "training wheels" mode, the best setup is often:

> Autopay the statement balance, plus alerts for large purchases, plus a low credit limit.

That combination creates a card that behaves like a debit card with extra protections and a reporting history.

Why minimum payments are so tempting (and so dangerous)

Minimum payments feel friendly because they're small. They're designed to be survivable even when money is tight. That's the point. The problem is what they do to time.

When you only pay the minimum, you're usually paying mostly interest and barely touching the principal. It's like trying to drain a bathtub with a teaspoon while the faucet is still on.

Parents can make this real without lecturing by connecting it to something teens already understand: subscriptions. A minimum payment can turn a one-time purchase into a subscription you didn't sign up for:

> "You're not paying for the headphones anymore. You're paying the monthly 'I still owe' membership."

There are situations where paying minimums temporarily is unavoidable. Life happens. The core lesson is to treat minimum payments like pain medicine: useful in an emergency, dangerous as a lifestyle.

The ``APR menu": one card can have multiple interest rates

Many teens assume an APR is *the* APR. One number. One price.

In reality, many cards have an APR menu:

- a purchase APR,
- a balance transfer APR (sometimes promotional),

- a cash advance APR (often higher, often starts immediately),
- and sometimes a penalty APR that can kick in under certain conditions.

This is one of those adult-life details that feels like a trap only because it's rarely explained plainly.

A clean way to say it: *a credit card is not one product; it's a toolbox. Different tools have different costs.* Purchases are usually the least painful if paid in full. Cash advances are often the "do not touch" button.

Cash advances: the ``emergency" that becomes an expensive habit

If your teen is going to have a card, they should know what a cash advance is before they ever see one in the wild.

A cash advance is using the credit card to get cash (from an ATM, for example). It often comes with:

- an upfront fee,
- a higher APR than purchases,
- and no grace period (interest starts immediately).

That's a triple whammy: fee, higher rate, and instant interest. It's one of the most expensive ways to borrow short-term. Teens don't need all the fine print; they need the rule:

> "Using a credit card to get cash is like buying cash with a markup."

If your teen is ever truly stuck—no money, no ride, no safe way home—you want them to call you, not discover the ATM option at 11:47 p.m.

Fees: the silent budget assassin

Interest is loud. Fees are sneaky. A teen can do everything "almost right" and still lose money to fees:

- a late fee because they paid one day after the due date,
- a returned payment fee because the checking account didn't have enough,
- a foreign transaction fee on a school trip,
- an annual fee they forgot was there.

A practical family move is to make the statement a short monthly ritual. Not an interrogation. A quick review:

> "Any interest? Any fees? Any charges you don't recognize? Anything you'd do differently next month?"

That's how you teach financial awareness without turning money into a shame topic.

Credit utilization: the gas gauge your kid doesn't know they're broadcasting

Credit utilization sounds technical, but teens understand it instantly when you translate it into a gas gauge.

Utilization is how much of your available credit you're using. If a card has a $500 limit and you're carrying a $250 balance when the statement closes, that's 50% utilization.

Why it matters: lenders and scoring models often interpret high utilization as a sign of stress—like someone driving with the fuel light on all the time. You might be fine. But it looks risky from the outside.

This creates a surprising lesson: you can be responsible and still look chaotic.

- You can pay on time every month and still have a lower score if your utilization is consistently high.
- You can also lower utilization by paying early (before the statement closes), not just by spending less.

This is a perfect "adulting" skill for teens because it's not about deprivation. It's about timing.

A simple rule of thumb many people use: try to keep utilization under about 30% on revolving accounts. Not because 31% is bad luck, but because lower tends to look safer. If your teen's first limit is tiny, even normal spending can blow past that, which is why early payment matters.

Case study: the straight-A kid who accidentally became a revolver

Here's a story that surprises parents because it doesn't involve a reckless teenager.

Jada is 17. Great student. Busy schedule. She gets a starter card with a $300 limit to use for gas and lunches. The first month is fine. The second month she buys a dress for a school event—$120—planning to pay it off when she gets paid from her part-time job.

Then life stacks up. A busy week. A math test. A shift swap. She pays *something* before the due date. She thinks she's being responsible. She is, in the sense that she's trying. But the statement balance isn't paid in full, so interest begins. Next month she's still carrying part of the dress, plus new gas charges, plus lunch charges. Her "small card" starts feeling tight. She bumps into the limit, pays it down, bumps again.

Within three months, Jada isn't using the card as a payment tool. She's using it as a float. Not because she's careless, but because her system wasn't designed for her life.

What fixes this isn't a lecture. It's mechanics:

- Autopay the statement balance (or a fixed payoff amount if cash flow is uneven),
- A weekly mini-payment (every Friday, pay whatever is on the card),
- Alerts when the balance hits, say, $150,
- A slightly higher limit *only if* spending controls are working (counterintuitive, but tiny limits can force high utilization and constant "limit stress").

The point of the story for a teen is liberating: *good intentions don't beat bad systems.* Build the system.

Training wheels: how to set up a first card without drama

Parents often imagine two extremes: either the teen shouldn't have a card at all, or the teen should have full freedom to "learn the hard way." There's a middle path that teaches skill without gambling with your child's future.

A strong training-wheels setup usually includes:

- *A very specific job for the card.* One category only at first: gas, a single subscription, or school lunches. Not "everything."
- *A low limit.* Low enough that a mistake is survivable, high enough that normal use doesn't constantly max it out.
- *Autopay.* Ideally statement balance, otherwise a fixed amount that reliably pays it off within a short window.
- *Notifications.* Purchase alerts, due-date alerts, and a balance threshold alert.

Then give the teen an identity that matches the goal:

> "You're practicing being the kind of person who uses credit without paying interest."

That framing is powerful. It's not "don't mess up." It's "this is what you're becoming."

Disputes and fraud: one reason credit cards can be safer than debit

One of the fairest arguments *for* credit cards is that they often provide better protections than debit cards for disputes and fraud. If a debit card is compromised, the money leaves your bank account. If a credit card is compromised, it's the bank's money on the line while the dispute is investigated.

Teens don't need a law lecture; they need two habits:

- Check transactions regularly (a quick weekly scan).
- Report anything weird immediately (small fraud becomes big fraud when ignored).

This can be a quiet confidence builder: credit isn't just temptation. It's also protection when used with awareness.

A simple household language that prevents misunderstandings

Families fight about money because they don't share definitions. A teen hears "pay your card" and thinks it means "send something." A parent means "pay it off."

Try adopting shared phrases:

- *"Pay the statement."* (Means pay the statement balance in full.)
- *"Pay the minimum."* (Means emergency mode; not a normal month.)
- *"Float."* (Means spending this month with next month's money.)

When everyone uses the same words, you can coach without nagging. It becomes operations, not morality.

Actionable takeaway: the two-rule system that keeps cards boring (in a good way)

If you want credit cards to stay a payment tool in your household, give your teen two rules that fit on a sticky note:

> *Rule 1: Pay the statement balance in full every month.*
> *Rule 2: If you can't do Rule 1, stop using the card until you're back to zero.*

Then set up the guardrails today:

- Turn on autopay (statement balance if possible).
- Turn on transaction alerts.
- Pick one category the card is allowed to cover for the first 90 days.
- Choose a weekly "card sweep" day where the balance gets paid down and the app gets checked.

A credit card used this way becomes wonderfully boring: a tool that builds reliability, protects purchases, and never charges rent for the privilege of using it.

6.3 Demystifying Credit Scores

Your kid can do everything right in plain sight and still get rejected in private.

That's the first thing to understand about credit scores: they don't grade character, kindness, or work ethic. They grade *borrower behavior as seen through a narrow window*. And because that window is narrow, it can produce outcomes that feel personal even when they're purely mechanical.

A credit score is best understood as a trust score—a prediction, based on past patterns, of how likely someone is to pay a bill on time. It's not a net-worth score. It's not a "how responsible are you as a human" score. It's closer to the kind of rating you get on a rideshare app: a blunt number meant to help strangers decide how risky it is to do business with you.

That framing alone can lower the temperature in a family. It replaces shame with strategy.

The invisible résumé your teen starts building before they realize it

Most teenagers treat "credit" like something future adults worry about, along with oil changes and colonoscopy reminders. Then the first adult milestone arrives: an apartment application, a first car insurance quote, a cell phone plan, a utility deposit, sometimes even a job screening. Suddenly someone is asking a question your teen didn't know was being asked:

> "Can we trust you to pay on time, consistently, without drama?"

A credit report is the résumé. A credit score is the quick summary at the top.

The frustrating part is that the résumé can be thin at exactly the moment your teen needs it. A young adult can have plenty of income and still look risky on paper because there simply isn't much history yet. That's why *starting early* matters—not with big debt, but with small, controlled proof of reliability.

There isn't one credit score

Most people talk about "my credit score" as if it's a single permanent number, like a blood type. In reality, you can have multiple scores at the same time, and they can differ depending on:

- which scoring model is used (there are several),
- which credit bureau's report the model pulls from (there are three major ones),
- and the day the score is calculated (because balances and updates change).

This is a relief when your teen panics after seeing a number online. One score isn't a life sentence. It's a snapshot taken with a particular camera.

If you want a calm, practical way to say it:

> "Different lenders use different recipes. You're not chasing one perfect number. You're building a pattern that looks good from any angle."

What a score is really measuring: risk, not worth

The money world has a habit of turning predictions into judgments. A credit score is fundamentally a risk score. It exists because lenders want an efficient way to estimate the chance they won't get paid back on time.

That creates a surprisingly healthy lesson for teens: you can be "good with money" and still have a mediocre score if you've never used credit. And you can have a good score while still being broke if you've used credit carefully and consistently.

So the goal isn't to worship a number. The goal is to build *options*. A solid score makes life cheaper and smoother: lower interest rates, fewer deposits, more approvals, more flexibility. It's like having a strong immune system. You don't notice it every day, but when something stressful happens, you're grateful it's there.

The big levers (and why two matter most for teens)

Credit scoring models look at several categories of information. The exact math is proprietary, but the broad ingredients are well known. A commonly cited breakdown looks roughly like this:

- *Payment history* (largest factor): do you pay on time?
- *Amounts owed / utilization*: how much of your available credit are you using?
- *Length of credit history*: how long have your accounts been open?
- *New credit*: recent applications and new accounts.
- *Credit mix*: experience with different types of credit.

For teens and young adults, two levers dominate because they're both powerful and controllable:

- *Payment history*: never miss a due date.

- *Credit age*: start early with something small and safe.

Notice what's *not* on that list: income, grades, how hard you work, or whether you "feel responsible." The system only sees what's reported.

Payment history: one late payment can be louder than twenty on-time ones

Payment history is the closest thing credit scoring has to a moral compass, but it's really about predictability. Lenders love boring. They love "always on time."

A late payment is a sharp signal because it suggests stress, disorganization, or both. For a teen, the lesson is not "never make mistakes." It's "protect your future self from avoidable mistakes."

That's why autopay isn't just a convenience—it's identity protection. When you set autopay for at least the minimum, you're building a moat around the most important category.

Then you add the second layer: make paying the statement balance the default habit whenever possible. That keeps the card from turning into expensive debt and prevents the "I'll deal with it later" spiral.

If you want a quick family mantra that sticks:

> "Late is expensive. Not just today—later."

Utilization: the score's weird obsession with how close you are to the limit

Utilization is the percentage of available revolving credit you're using. If your teen has a $500 limit and the statement closes with a $250 balance, that's 50% utilization.

Here's why this matters so much: high utilization can make a person look maxed out, even if they pay in full every month. Scoring models tend to interpret heavy use of available credit as a sign of potential overextension.

Many experts use a rough rule of thumb: keep utilization at or below about 30%. It's not a magic line, but it's a helpful target.

This is where teens get accidentally tripped up, because starter limits are often low. A teen can spend $180 on gas and groceries, completely reasonable, and suddenly be at 60% utilization.

The fix isn't always "spend less." It can be "pay earlier." You can make an extra payment before the statement closes so the reported balance is smaller. In other words: you can manage utilization with timing, not just with deprivation.

A clean script for a teen:

> "Your limit is like a stage. If you spend near the edge, the audience thinks you might fall—even if you never do."

Credit age: the boring superpower

Credit age rewards patience. The longer you've had accounts open and in good standing, the more stable you look. That's why a teen who starts at 18 with a simple card and uses it responsibly can have a meaningful advantage by 21.

This is also why closing old accounts can sometimes hurt. People assume closing a card is always "responsible." Sometimes it is. But it can shorten your average credit age and reduce available credit, pushing utilization higher.

Your teen doesn't need to obsess over this. They need one idea: *start early, keep it clean, and let time do the heavy lifting.*

The myth that won't die: ``carry a balance to build credit''

This one is everywhere, and it costs families real money.

You do *not* need to carry a balance and pay interest to build credit. You build credit by using credit and paying it on time. The system rewards reliable repayment behavior; it doesn't require you to pay interest as an entry fee.

Think of it like this: if you rent a car and return it on time with a full tank, they trust you. They don't trust you more because you returned it late and paid a penalty.

A teen-friendly way to say it:

> "Interest is not a membership fee for a good score.
> It's a penalty for not paying the bill."

Case study: two 19-year-olds, two very different ``trust scores''

Consider two freshmen, both working part-time.

Ava avoids credit entirely because she's heard it's dangerous. She uses a debit card and cash. She's careful. She never misses a payment on anything—but she doesn't have anything that reports payment history.

Noah has a starter credit card with a $500 limit. He uses it for one subscription and gas. He set autopay for the statement balance. Once a month, he checks the statement for mistakes and moves on with his life.

At 19, Noah may already have a score because he's producing data: on-time payments, low utilization, consistent behavior. Ava may have little or no score because she has no credit history.

Neither is a better person. But the system will treat them differently. When they apply for the same apartment at 21, Ava might be asked for a larger deposit or a co-signer, while Noah slides through more easily.

That's the point of building credit early: not to borrow more, but to be asked for *less proof* when life requires it.

How to help a teen build credit without turning them into a spender

Parents sometimes worry that "building credit" is code for "encouraging consumerism." It doesn't have to be.

The safest approach is to connect credit building to expenses your teen already has and already plans for. The card is not a shopping invitation; it's a reporting tool.

Here are "clean" ways to generate positive history:

- Put a small, predictable subscription on the card (music, cloud storage, etc.) and autopay it.
- Use the card for gas only, with a weekly payoff.
- If your teen is an authorized user on a parent's card, treat it like a learner's permit: clear rules, low limits, immediate reimbursement, and transparency.

The theme is always the same: *small charges, paid in full, on autopilot.*

Credit reports: the file behind the score (and why accuracy matters)

The score gets all the attention, but it's built from your credit report. The report is a record of accounts, limits, balances, and payment behavior. If the report is wrong, the score can be wrong.

Errors are more common than most people assume, and identity theft can show up first as a weird account you don't recognize.

A powerful family habit is to treat credit reports like a yearly physical: not because you expect bad news, but because catching something early saves pain later.

In the U.S., consumers are entitled to free credit reports from each of the major credit bureaus through the authorized site (AnnualCreditReport.com). A parent can make this a rite of passage around age 18:

> "Let's pull your reports, make sure they're clean, and learn what your name looks like in the financial system."

That experience demystifies the whole thing. It turns credit from spooky folklore into paperwork you can read.

A gentle warning about ``credit-building" products

If you search "build credit fast," you'll find plenty of products that promise shortcuts. Some are legitimate, some are overpriced, and some are simply unnecessary.

Your teen doesn't need complexity. A single low-limit card used responsibly, or a well-structured starter option, can do the job. Credit is more like fitness than like a test. There's no hack that beats consistency.

If your teen is being marketed a solution, ask two questions together:

> "What does this cost?" and "What problem does it solve that a simple card doesn't?"

If the answers are fuzzy, walk away.

Actionable takeaway: a simple ``Credit Score Playbook" for ages 16--22

Here's a straightforward playbook you can use at the kitchen table without turning it into a lecture:

- *One job for the card.* Pick a category (gas or one subscription) so spending doesn't sprawl.

- *Autopay at least the minimum.* Non-negotiable protection against late payments.
- *Aim for low utilization.* Keep the statement balance under about 30% of the limit when possible; if the limit is tiny, make a payment before the statement closes.
- *Never pay interest on purpose.* Pay the statement balance in full whenever you can.
- *Once a year, check credit reports.* Make sure the story being told about your teen is true.

If your teen does just those things, their score will tend to rise the way a good reputation rises: slowly, quietly, and powerfully—until one day they need it, and the door opens without an argument.

6.4 Avoiding Debt Traps

Debt traps don't usually look like traps. They look like help.

They show up when your teen is tired, stressed, short on time, or embarrassed to ask. The pitch is always the same, whether it comes from a neon-lit storefront, a slick app, or a "pre-approved" offer in the mail: *We can make this problem go away right now.* And that's exactly why the products work. They sell relief, not math.

If you want to protect your kid, you don't need to turn them into a cynic. You need to turn them into a good detective. A detective doesn't assume every offer is evil. A detective asks the two questions traps hate most:

> "What does this really cost?" and "What happens if I'm late?"

The anatomy of a debt trap: fast money, tight deadlines, expensive consequences

Most healthy borrowing is slow and boring. You apply, you qualify, you sign, you repay on a schedule that assumes you're human. Debt traps flip that.

A classic trap has three features:

- *Speed:* money today, with almost no friction.
- *Short runway:* repayment due quickly (often in one lump sum).
- *A penalty engine:* fees and interest that explode when you can't repay exactly on time.

That combination doesn't just profit from borrowers who struggle. It *depends* on them. If everyone repaid easily on the first due date, the product wouldn't be very profitable. The business model often assumes repeat borrowing, rollovers, reborrowing, and fees that stack.

A teen-friendly translation is:

> "If the plan only works when everything goes perfectly, it's not a plan. It's a bet."

Payday loans: the ``two-week" loan that can last months

Payday loans are often marketed as tiny, temporary bridges: borrow a little now, repay on your next paycheck. The problem is that the bridge is built over a canyon, and the toll is outrageous.

Here's the mechanism that makes payday loans so dangerous:

- the loan is due quickly (often in a single payment),
- many borrowers can't repay the full amount plus fees when the due date arrives,
- so they roll it over, renew, or take another loan to cover the first.

The trap isn't just the high rate. It's the *mismatch* between a short due date and real life. Rent, food, gas, and phone bills don't pause because a payday loan is due.

When you explain payday loans to a teen, you don't have to lead with a scary APR number (though it's worth mentioning they can be astronomically high). Lead with the structure:

> "A payday loan is usually one big payment due fast. If you can't make that payment, you don't get a second chance—you get a more expensive loan."

Auto title loans: borrowing against your ability to get to work

An auto title loan is the kind of product that sounds reasonable if you don't look closely: you own a car, you borrow against it, you repay. The brutal part is what's being used as collateral.

For many working families, a car isn't a luxury. It's access to income. It's how you get to work, school, childcare, medical appointments. Borrowing against the title means one mistake can risk your transportation. And once transportation is in danger, income becomes unstable. That's when financial problems start multiplying.

A teen doesn't need the full legal history of title lending. They need the core warning:

> "Never risk losing the thing that lets you earn money in the first place."

If borrowing could lead to losing the car, the loan isn't just expensive—it's destabilizing. It's like pulling bricks out of your own foundation to patch a roof leak.

Buy Now, Pay Later (BNPL): the nice-looking trap that feels like budgeting

Buy Now, Pay Later is the most modern-looking debt trap because it wears the costume of responsibility.

The pitch is soothing: four payments, no interest, easy approval. It doesn't even *feel* like debt because it often doesn't look like a credit card. It looks like a checkout option.

BNPL can be genuinely useful for someone who is disciplined, has stable cash flow, and wants a short payment plan for a purchase they already planned for. The trap appears when:

- multiple BNPL plans stack up,
- payments are scheduled across different dates,
- and the borrower loses track of how many "small" obligations are already committed.

The danger is not one BNPL plan. The danger is ten small plans that add up to a big monthly drain.

A great line for teens is:

> "BNPL turns one shopping decision into a calendar of future bills."

That calendar effect is where overextension happens. The present moment feels cheap; the next month becomes crowded.

The ``small fee" trick: when a fee is really an APR in disguise

Predatory offers often avoid talking about APR because APR sounds scary. Instead, they talk about "flat fees," "service charges," "tips," "expedite costs," or "membership."

Here's the problem: a small fee can be an enormous APR if the loan term is short.

If your teen borrows $100 and pays a $15 fee to borrow it for two weeks, that doesn't sound like "390% APR" at first glance. It sounds like $15. But the shorter the time, the more that fee behaves like a sky-high interest rate.

You don't have to do complex math with your teen every time. You can teach the smell test:

> "If you're paying a fee just to access your own next paycheck, the product is expensive by design."

And another one:

> "If the cost is explained in dollars but the time is explained in days, slow down. That's where big APRs hide."

Deferred interest: the retail trap that punishes ``almost"

Some store credit offers are structured like a dare: "No interest if paid in full in 12 months."

That can be fine if the borrower truly pays it off by the deadline. But "deferred interest" is different from "0% APR." With deferred interest, if you miss the payoff deadline by even a little, you may owe interest that was quietly accumulating in the background, sometimes retroactively.

The trap is psychological: people hear "no interest" and relax. They make smaller payments. They assume they're safe. Then the deadline passes and the bill changes shape.

A parent script that keeps it simple:

> "If it says 'paid in full by,' treat that date like a final exam. Don't aim to pass. Aim to finish early."

If your teen ever uses a promo like this, the only safe way is to divide the purchase by the number of months and pay *more than* that amount, so there's a cushion. Life happens. Cushion is the difference between "deal" and "disaster."

Minimum payments: the trap hiding in plain sight

Not all traps are exotic products. Some are normal features that become dangerous through habit.

Minimum payments exist to keep an account from being delinquent. They do not exist to make debt cheap. When a teen learns "minimum payment" as the default, they're training themselves to live perpetually one step behind.

The trap here is subtle: nothing explodes. There's no scary phone call. No neon sign. Just a slow leak of money into interest, month after month, while the balance refuses to disappear.

A useful phrase to teach is:

> "Minimum is the least they'll accept, not the amount you should pay."

A real-world scenario: the ``fix my paycheck" spiral

Imagine your teen, Maya, has a part-time job and a checking account. She miscalculates her balance because two subscriptions hit earlier than usual, and her account goes negative the day before payday. She's embarrassed and wants to solve it quietly.

An app offers her $75 "early" for a small fee, and she takes it. Payday arrives, but her paycheck is smaller than expected because she worked fewer hours. The app takes its money back. Now she's short again. She takes another advance, plus another fee.

Nothing here feels like a loan at first. It feels like smoothing out timing. But the fees repeat. The "advance" becomes part of her monthly budget even though it's supposed to be occasional. That's the spiral: the product becomes a permanent patch for a recurring cash-flow problem.

A parent doesn't need to scold Maya for being careless. Maya didn't need shame; she needed a better tool:

- a small buffer in checking (even $100 makes a huge difference),

- a rule that subscriptions only come from one "bills" account,
- and a habit of checking balance before spending, not after.

The deeper lesson for your teen is empowering:

> "If you're borrowing to fix timing every month, it's not timing. It's a budget gap."

Budget gaps don't get solved with more borrowing. They get solved with smaller fixed expenses, more income, or a buffer.

The Red Flag Checklist: how teens can spot a predatory offer in 60 seconds

You're not trying to train a teen to read contracts for fun. You're trying to give them a fast internal alarm system. Here are red flags worth teaching as a quick mental scan:

- *"No credit check!"* can mean "we're charging you because we assume risk."
- *Repayment is due very soon* (two weeks, one month, or a single lump sum).
- *Costs are described as fees, not interest* (especially on short-term loans).
- *The offer is framed as secrecy-friendly* ("no one has to know," "instant approval," "no questions").
- *Collateral is essential to your life* (car title, access to your bank account).

- *The salesperson pushes urgency* ("today only," "limited time," "act now").

If your teen can recognize those patterns, they'll avoid a lot of pain without needing to memorize APR tables.

What to do instead: emergency options that don't eat your future

Avoiding traps is easier when your teen has a short list of safer alternatives. The goal is to reduce the chance that "panic borrowing" becomes their default.

Depending on your family and your teen's situation, that list might include:

- calling you (or another trusted adult) before borrowing,
- asking a biller for a due-date change or payment plan,
- using a small personal buffer fund (even $200 can stop a spiral),
- picking up an extra shift before taking a loan,
- selling a non-essential item instead of borrowing against essentials.

The strongest alternative is emotional, not financial: giving your teen permission to ask for help early. Traps thrive on silence. They shrink when a teen can say, "I messed up, and I want to fix it."

Actionable takeaway: a family ``pause rule" that blocks most traps

Here's a rule that works because it respects real life while protecting your teen from high-cost decisions made under pressure:

> *If a loan promises instant relief, we wait 24 hours and talk to one human before accepting it.*

During that 24-hour pause, your teen answers three questions (in a text to you, a note on their phone, or out loud at the kitchen counter):

- What problem am I trying to solve right now?
- What is the total payback amount and the exact due date?
- If I'm late, what happens—fees, higher rates, losing my car, losing access to my bank account?

That pause rule doesn't require perfect self-control. It creates a speed bump. And for debt traps, one speed bump is often the difference between a small mistake and a long, expensive story your kid has to drag behind them for years.

6.5 Paying for Higher Education

The most expensive sentence in the college process is often said with love: *"Don't worry about the cost—we'll figure it out."*

Families mean, *We believe in you.* Colleges hear, *Price sensitivity is low.* And loans are happy to translate optimism into paperwork.

College can be a phenomenal investment. It can also be a financial ankle weight that changes where a young adult can live, what jobs they can take, how quickly they can leave a bad boss, whether they can start a business, and how soon they can buy a home. The difference is rarely intelligence. It's usually *price paid* and *plan for paying.*

A useful way to talk about college with teens is to take the emotion out of the loan part without taking the meaning out of the dream. College is not just a place you go. It's a project you finance. And any financed project has two numbers that matter more than the brochure:

- *What is the total cost to get the credential?*
- *What is the expected payoff in earnings and options afterward?*

That's ROI thinking—return on investment—applied to a real human life.

The budget talk before the application talk

Most families do this backward. They tour campuses, fall in love with a vibe, imagine dorm life, and only then ask, "So... how much is it?"

By that point, the conversation is no longer about money. It's about grief. Saying no feels like canceling a future.

Flip the order. Start with the budget, then shop within it. This is not being "practical" in a joyless way. It's choosing joy that doesn't come with a monthly bill for the next decade.

A parent script that lands without sounding harsh:

> "We're not deciding where you *can* go. We're deciding what we're willing to *finance*."

That sentence is a gift. It separates worth from price.

Sticker price is not the price

Colleges love sticker price because it makes discounts feel generous. Families love sticker price because it's a simple number to compare. Both are wrong to focus on.

The number that matters is *net price*: the cost after grants and scholarships (free money) reduce the bill. Loans don't reduce net price—they just change who pays and when.

Many colleges are required to provide a net price calculator on their website. Use it early, even if the estimates are rough. A rough number now beats a precise shock later.

A teen-friendly translation:

> "Sticker price is the list price. Net price is what your family actually pays."

When you teach your kid to ask for net price, you're teaching them a life skill that goes far beyond college: never negotiate from the advertised number.

The strategy: a funding funnel instead of a debt buffet

When families are stressed, they often fund college like they fund a surprise car repair: swipe, borrow, figure it out later. That approach is how student debt becomes a fog.

A better approach is an order of operations—a funnel that starts wide with the best money and narrows down to the riskiest money.

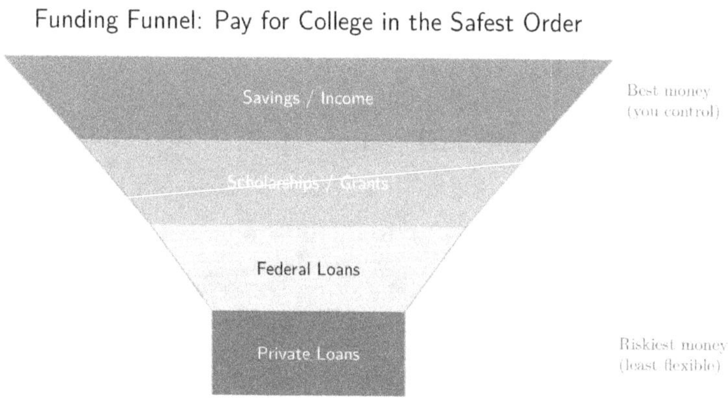

The funnel is simple on purpose. It prevents one of the most common family mistakes: borrowing first because it's easiest, then trying to "make it up" later. That's backwards.

Layer 1: Savings and income (the money with the fewest strings attached)

The top of the funnel is money you already have or can earn without signing away future options:

- savings accounts,

- 529 plans (if you have them),
- student income from work-study or part-time work,
- parent contributions that don't require loans.

This layer matters for a deeper reason than "avoiding debt." It keeps your kid flexible. A graduate with lower required payments can take an internship, choose a lower-paying first job with strong growth, move for a better opportunity, or leave a toxic workplace. A graduate with a heavy loan payment becomes easy to trap.

If your teen works during school, the goal isn't to martyr them. It's to create *ownership*. Even a small contribution changes behavior. Students who are investing their own effort often shop differently, register for classes differently, and take the degree more seriously.

Layer 2: Scholarships and grants (free money, but with a schedule)

Scholarships and grants are the closest thing to a financial cheat code, but they have rules. Scholarships can require maintaining a GPA, staying in a particular major, or meeting renewal criteria. Grants can depend on financial circumstances.

The practical lesson for teens is that free money is still a contract:

> "If your scholarship requires a 3.0 and you're sitting at a 3.1, that's not a fun fact. That's your job."

Parents can help by building a simple annual ritual: once per term, review scholarship requirements the same way you'd review a lease. Not dramatic, just disciplined.

Layer 3: Federal student loans (not painless, but usually more forgiving)

When borrowing is necessary, federal student loans tend to be the safer category because they come with borrower protections and standardized terms. They can still be serious debt, but they're typically less predatory than many private alternatives.

A key concept teens can understand: federal loans come in different flavors, including *subsidized* and *unsubsidized*. The practical difference is about who pays the interest while the student is in school and during certain periods. Subsidized loans are generally more favorable if the student qualifies because the interest burden is reduced during school.

Even here, the best habit is restraint:

> "Just because you're approved for a loan doesn't mean you should take the full amount."

Loan offers can feel like awards. They are not awards. They're prices.

Layer 4: Private loans (the narrow red bottom for a reason)

Private student loans are at the bottom because they tend to be the least flexible and can carry higher costs and fewer protections. They may require a co-signer (often a parent), which means your family is sharing risk long after graduation.

The hard truth: private loans can follow a kid into adulthood like a shadow that demands cash every month regardless of whether the degree delivered what was promised.

If private loans enter the conversation, treat it like a board meeting, not a casual click:

- Compare fixed vs. variable rates.
- Know exactly when repayment begins.
- Understand co-signer obligations.
- Ask what happens in hardship (and get answers in writing).

The strongest guidance you can give a teen is values-based and practical:

> "Your future is not something to finance with the most fragile kind of debt."

The Borrowing Limits Mindset: don't borrow more than the first-year salary you expect

If you want one rule that is strict enough to matter and simple enough to remember, use this:

> *Don't borrow more in total than the salary you realistically expect in your first year after graduation.*

This isn't a perfect rule. Some fields have lower starting salaries but strong long-term trajectories. Some careers require graduate school. Some students live at home, lowering costs. But as a baseline, it forces the right conversation: *loans must fit inside the life they're supposed to support.*

A teenager can grasp this instantly because it connects debt to paycheck reality. It's the opposite of fantasy math.

ROI without coldness: what are you buying, exactly?

ROI talk can sound like you're reducing education to dollars. You're not. You're refusing to let a lender price your child's future based on vibes.

A degree can deliver value in different forms:

- higher earnings,
- access to certain professions (licensure),
- networks and internships,
- a structured environment for growth.

The financial question is: are you paying a price that matches the likely benefit?

A parent can keep this human by making it about freedom:

> "Education should expand your options. Debt should not shrink them."

Case study: two offers, same dream, different math

Suppose your teen, Jordan, wants to study nursing. Two schools accept them.

School A is a private college with a beautiful campus. After scholarships, the net price is $38,000 per year. Jordan would need to borrow roughly $25,000 per year to make it work.

School B is a public university with a strong program. After scholarships, the net price is $18,000 per year. Jordan would need to borrow $5,000–$8,000 per year.

Jordan likes School A more. Many teens would. The campus feels like a movie. The brochures are glossy. The tour guides are trained to help you imagine yourself there.

Now the adult question: what's the total debt at graduation?

Four years of borrowing $25,000 per year can put Jordan near $100,000 in debt (before you even consider interest). Four years of borrowing $8,000 per year might land closer to $32,000.

Those two numbers create two different lives:

- One life requires a high monthly payment that limits choices right out of school.
- The other life leaves room for moving, saving, taking a slightly lower-paying first job with a great mentor, or working part-time while finishing a specialization.

Jordan can still choose School A. But now it's an adult choice, not a romantic one. And that's the real win: your teen learns to make big decisions with both heart and calculator.

The ``net price + outcomes" habit: use data like a grown-up

When families compare colleges, they often compare feelings: campus vibe, dorms, sports, how friendly the tour guide was. Those things matter, but they're not the financial decision.

A sturdier method is a two-column comparison:

- *Net price* (what you actually pay),
- *Outcomes* (graduation rates, typical debt, typical earnings).

Tools like the College Scorecard exist for exactly this reason: to make outcomes easier to compare across schools. The numbers are not perfect, and different metrics may come from different years, but they're far better than guessing. Teach your teen to treat "typical debt" and "typical earnings" as real features, like lab facilities or class sizes.

A line that keeps it grounded:

> "Every school sells a story. Data helps you check whether the story usually comes true."

A note on timelines: five years is not the same as four

One of the quiet debt multipliers is time-to-degree. If a student takes five or six years to finish, costs rise even if tuition stays the same:

- extra semesters of tuition and fees,
- extra housing and food,
- delayed full-time earnings.

Parents can help without becoming a micromanager by focusing on planning:

> "Before each term, let's confirm your classes keep you on track to graduate."

That habit protects ROI more effectively than arguing about lattes.

Actionable takeaway: the ``College Cost Conversation'' you can have in one hour

Here's a practical way to run a single, high-impact family meeting that can save tens of thousands of dollars and years of stress. Bring a notebook or a shared document and answer these questions together:

- What is our maximum *annual* out-of-pocket amount (without loans)?
- For each school: what is the estimated net price (after grants/scholarships)?
- If loans are needed: how much total debt would exist at graduation?
- What first-year salary range is realistic for this degree in the locations your teen would actually live?
- Does the total borrowing stay under that expected first-year salary?

If the numbers don't work, don't treat it as failure. Treat it as design. Adjust the plan: start at a community college, live at home, choose a lower-cost program, increase scholarship effort, or pick a school that delivers similar outcomes for a lower net price.

The best college funding plan isn't the one that looks impressive on decision day. It's the one that lets your kid graduate with momentum—a degree in hand and enough breathing room to build an adult life that isn't already financially cornered.

Chapter 7

Investing for the Future

If saving is storing your seeds in a jar, investing is planting them in the ground. It is the difference between preserving what you have and growing a forest that can shade your children's future. In this chapter, we will decode the scary jargon of Wall Street and reveal the simple, boring, yet incredibly effective strategies that turn modest savings into generational wealth. We aren't trying to beat the market; we're teaching our kids to join it for the long haul.

7.1 The Investing Mindset

A teenager opens a brokerage app for the first time, buys $200 worth of a big, famous company, and feels oddly proud—like they just planted a flag. Two days later the price is down $18. Their stomach drops. They didn't just "lose eighteen dollars." They lost something bigger: the comfortable belief that money stays where you put it.

Most adults aren't any different. We say we're afraid of "the market," but that's not quite true. What we're afraid of is the feeling of watching a number go down and not being able to make it stop. Investing asks you to tolerate a specific discomfort: uncertainty with your own money. It's personal. It's emotional. And it's exactly why the mindset matters more than the math in the beginning.

Here's the mental switch that unlocks investing: learn to tell the difference between *volatility* and *loss*.

Volatility is the normal up-and-down wobble of prices. It's the bumpy flight, not the plane crash. Loss—real loss—is when money is gone in a way you can't reasonably recover from: you sold during a panic and locked in the drop; you concentrated in something that failed; you needed the money at the exact wrong time; or inflation quietly shrank what your savings can buy. Volatility is a feeling. Permanent loss is a fact.

The market is not a polite escalator that carries your money upward. It's a crowded city street: noisy, unpredictable, occasionally chaotic, and full of people reacting to rumors, fear, excitement, headlines, and plain old impatience. Prices move because people move. When you understand that, the daily wiggles stop feeling like a verdict on your intelligence. They start looking like what they are: the admission price for long-term growth.

Why ``I'll Just Keep It in Cash" Feels Safe (and Isn't)

Cash has a superpower: it doesn't surprise you. A $20 bill is always a $20 bill. A savings account statement is calm. For parents teaching kids, calm feels like responsibility.

But "safe" has two meanings:

- *Safe from drama* (no scary dips on a chart).
- *Safe for your future self* (keeps up with the cost of life).

Cash is excellent at the first meaning and unreliable at the second. The enemy here isn't a villain you can point to. It's inflation: the slow rise in prices over time. Inflation is why your parents could buy movie tickets for a few dollars and why a "cheap car" is now a phrase people say with a laugh.

Inflation doesn't usually feel like theft because it doesn't happen all at once. It's more like a leak under the sink. You don't notice day-to-day, and then one day the cabinet floor is warped and you can't remember when it started. If you keep $10,000 in cash for decades, you may still have $10,000—but it will buy less life.

And here's the part that messes with our instincts: the "no drama" choice can be the one that quietly guarantees you fall behind.

Inflation: The Silent Killer

When families say, "We're just trying not to lose money," they often mean "We don't want our balance to go down." Inflation doesn't make your balance go down. It makes your *options* go down.

Think in purchasing power: not "How many dollars do I have?" but "How many groceries, rent payments, braces, laptops, or plane tickets does this represent?" If those prices rise and your cash doesn't, the gap is the loss. It's just wearing a disguise.

Even if inflation averages a modest rate over time, the compounding works against you. You don't need a scary headline for it to matter. You just need time.

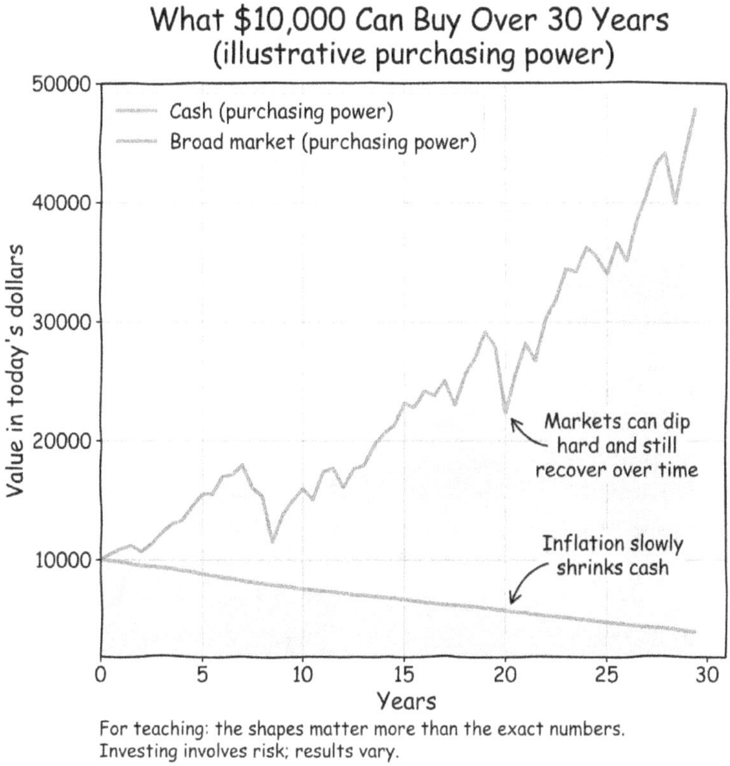

If your kid looks at that chart and says, "So the market line is better," that's a great opening. Then ask the more grown-up question: "Why doesn't everyone do it?" Their answer will usually be the honest one: "Because the green line is scary."

Exactly.

The Brain Hates Losing (Even When It's Small)

Humans aren't wired to think in 30-year timelines. We're wired to react to immediate threats. A paper cut gets more attention than a slow decline in cardiovascular health. A market drop feels like danger right now; inflation feels like a fog that might never touch you.

That wiring shows up in everyday investing mistakes:

- *Panic selling*: "I can't take this anymore," followed by selling when prices are low.
- *Performance chasing*: buying whatever has been going up lately because it feels "safe."
- *Checking too often*: turning a long-term plan into a daily referendum on your judgment.

Notice how none of those are math problems. They're nervous-system problems.

One of the best things you can teach a kid is that a red day on an investment app is not an emergency. It's weather. Sometimes it storms. You don't sell your house because it rained.

A Real-World Case: The Family Who Learned the Wrong Lesson

Consider a very common story. A family starts investing for the first time in their 30s. They put a chunk of savings into a stock fund. A few months later, the market drops sharply. The balance is down thousands. They feel embarrassed—like they did something reckless—and they pull the money out "until things calm down."

Years pass. They never quite feel like it's calm enough to go back in. They miss the recovery, and they miss the growth that usually follows recoveries. The lesson they carry forward isn't "markets are volatile." It's "investing is dangerous."

Now look at what actually happened. The market didn't take their money. Their decision did. The permanent loss wasn't the drop; it was selling during the drop.

This is why "I don't want to lose money" needs a follow-up question at your kitchen table: "Do you mean you don't want your balance to dip, or you don't want to permanently damage your future?"

That single question can change a family's financial trajectory.

Volatility Is a Feature, Not a Bug

It sounds strange, but volatility is part of why stocks have historically offered higher long-term returns than safer-feeling assets. If an investment were perfectly stable and perfectly rewarding, everyone would want it, and the price would rise until the future return wasn't so generous anymore.

A helpful way to explain this to a teen:

> A savings account is like mowing a lawn for a fixed $25. It's predictable.
> Stocks are like taking a job where some weeks you make $10 and some weeks you make $80, but over the summer you usually make more than mowing lawns. The trade is emotional comfort for long-term growth.

You're not teaching your child to love risk. You're teaching them to understand *which risk they're actually taking*.

The Two Kinds of Risk That Matter Most

If you want a kid-friendly framework that still holds up for adults, focus on these two:

1) *Time risk (needing the money too soon).*
Money for next year's car insurance or next semester's tuition doesn't belong in something that can drop 30% right before you need it. That's not being "conservative." That's being honest about the timeline.

2) *Concentration risk (betting too much on one thing).*
If all your money depends on one company, one industry, one crypto coin, or one "can't miss" idea, a single bad outcome can become permanent. Diversifying is not about being fancy; it's about refusing to let one storyline control your life.

When families get burned, it's often because of one of these two—not because they invested broadly and patiently.

A Practical Teaching Moment: The ``Two Buckets" Rule

When you're raising kids, you don't need a lecture to build good instincts. You need simple rules that match reality.

Try this: whenever your child earns, receives, or saves a meaningful amount, label it as going into one of two buckets.

- *Soon Money*: anything needed within the next 1–3 years (phone replacement, first car, summer program, emergency buffer). This bucket should be boring on purpose—cash or very safe options.
- *Later Money*: anything truly meant for far-off goals (a head start on retirement, a long-term gift, "future me").

> This bucket can handle the market's mood swings because it doesn't need to be touched in a hurry.

The magic here is psychological. Once "Later Money" is mentally separated, a downturn stops feeling like a threat to next month's life. It becomes a temporary markdown on something you plan to hold.

If your teen is old enough to have an app, give them a script for red days:

> "My timeline didn't change, so my plan doesn't change."

That's not stubbornness; it's discipline.

The Myth of the Perfect Moment

Kids are very alert to "getting tricked." They don't want to be the person who bought right before a crash. Adults feel the same and just hide it better.

So name it: the desire to wait for a perfect moment is normal—and usually expensive.

If you only invest when things feel calm and safe, you tend to buy when prices are already high. If you flee when things feel scary, you tend to sell when prices are already low. The market has a cruel sense of humor like that.

A steadier approach is to invest the same amount regularly, whether the market is cheerful or gloomy. Some months you buy at higher prices, some months at lower prices. Over time, you build a habit instead of a hunch. For a family, that habit is the real victory because it doesn't require anyone to be a hero.

What to Say When Your Kid Asks, ``Can I Lose Everything?"

This is the question underneath all the other questions, and it deserves a straight answer.

If you put all your money into one fragile thing, yes, you can lose nearly everything. Companies fail. Scams happen. Fads burn out.

If you spread money across a broad slice of the economy and give it time, the realistic risk changes shape. Your account value will swing—sometimes dramatically—but the odds of "everything goes to zero" become very different because you're not betting on one outcome. You're buying participation in thousands of businesses that sell groceries, medicine, electricity, software, sneakers, and everything else people keep buying year after year.

That doesn't make it guaranteed. It makes it *structurally more resilient*.

And there's a second honest answer: you can "lose everything" in slow motion by never investing at all and letting inflation take the wheel. It just won't feel dramatic enough to warn you.

A Parent's Secret Weapon: Make the Conversation About Behavior, Not Bravery

A lot of investing advice accidentally rewards swagger: "Be fearless," "Be bold," "Go all-in." That's fun to watch in a movie and disastrous as a family strategy.

The strongest investing mindset is quieter:

- You respect uncertainty instead of pretending it doesn't exist.
- You build rules that protect you from your own worst impulses.
- You measure success by consistency, not by winning a single year.

That's a powerful lesson for kids because it transfers beyond money. It teaches them to separate feelings from decisions, to tolerate discomfort in service of a long goal, and to avoid confusing luck with skill.

Actionable Takeaway: A 10-Minute ``Volatility Practice" You Can Do This Week

Pick a single broad-market fund or index your family recognizes (the kind that represents "lots of companies," not a single name). Then do this at the dinner table:

1. Look up a long-term chart (10+ years).
2. Point to three scary drops and ask: "If you sold here, what would you be feeling? If you held, what happened later?"
3. Decide on one family rule and write it down somewhere visible:
   "We don't sell Later Money because of headlines."

If your child can learn—early—that a temporary drop is not the same thing as a permanent loss, they'll have something rarer than a hot stock tip: they'll have emotional control around money. That's the skill that keeps good plans alive long enough to work.

7.2 The Eighth Wonder: Compound Growth

A lot of teenagers think investing is something you do *after* you become rich—like buying a nice watch, or upgrading to first class. Adults often think the same thing, just with better vocabulary. They imagine investing is powered by big paychecks.

Compound growth has a different opinion. It doesn't care how impressive your income sounds at a party. It cares about one thing: *time*.

That's the first mental twist worth giving your kid: the most valuable thing a young investor has isn't cash. It's years. A fourteen-year-old with a part-time job has something a forty-year-old surgeon can't buy back: decades of compounding.

The Not-So-Secret Trick: Money That Earns Money (That Earns Money)

Compound growth is the snowball effect of investing. You don't just earn returns on the money you put in. You earn returns on the returns.

Say you invest $100 and it grows 10% in a year. Now you have $110. If it grows another 10% the next year, you don't earn $10 again-you earn $11, because the "base" got bigger. That extra dollar is the whole game. It looks tiny early on, which is why people ignore it. Then it becomes enormous later, which is why people wish they hadn't.

This is also why the most dangerous sentence in personal finance is, "I'll start later, when I have more." Starting later is like arriving at a planting season after everyone else already put seeds in the ground. You can still grow a garden—but you'll spend the rest of the summer staring at your neighbor's tomatoes.

A Family-Friendly Definition (That's Actually Accurate)

When kids hear "compound interest," they picture a bank account. Investing is broader than that, but the idea transfers cleanly.

Compound growth means your money can grow in a loop:

> You invest. It grows. You keep the growth invested.
> Now the growth can grow.

The loop is powered by reinvestment. That reinvestment happens automatically in many investment accounts if you choose it (for example, reinvesting dividends). Or it happens simply because you didn't cash out. Either way, the key behavior is the same: *don't interrupt the loop.*

The ``Rule of 72": A Little Math That Feels Like a Magic Trick

There's a quick way to estimate how long it takes money to double: the Rule of 72.

Take 72 and divide it by the annual return (as a percent). The result is roughly the number of years to double.

- At 8% per year: $72 \div 8 \approx 9$ years to double.

- At 6% per year: $72 \div 6 = 12$ years to double.
- At 3% per year: $72 \div 3 = 24$ years to double.

This is a fantastic dinner-table tool because it turns investing from "mystery" into "timing." It also teaches a subtler lesson: returns matter, but the number of *doublings* matters more. A young person isn't just trying to "make money." They're trying to stack doublings.

Chloe vs. Sam: The Story That Makes Time Feel Real

Here's a thought experiment that routinely makes smart adults blink.

Chloe starts investing at 15. She puts in $100 a month for 10 years (from 15 to 25). Then she stops completely—no more contributions. She leaves the account alone and lets it grow until she's 65.

Sam starts investing at 25. He puts in $100 a month from 25 all the way to 65 (40 years). He's consistent, responsible, and never misses a year.

Same monthly amount. Sam invests for four times as long. It feels obvious who should end up with more.

But compounding doesn't reward effort the way we expect. It rewards *early time in the market*. Chloe's money gets to compound for 50 years. Sam's earliest dollars compound for 40 years, but his later dollars have much less time—some contributions only have a few years to grow.

To make the comparison concrete, assume an average 7% annual return, compounded monthly. (That number is not a promise; it's a round, commonly used long-term illustration.) Now look at the result.

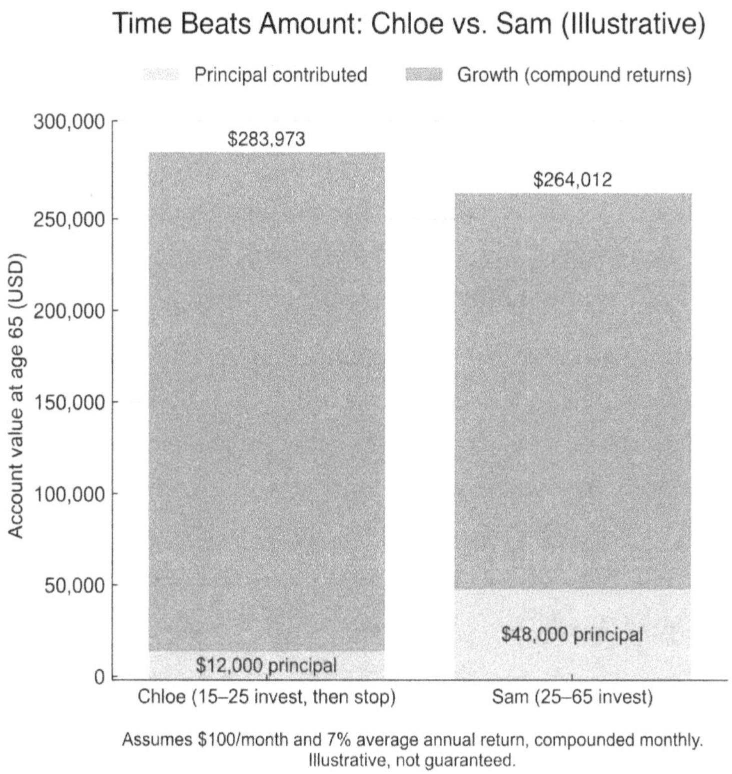

In this illustration, Chloe often ends up with *as much as or more than* Sam, despite contributing far less total money. That is not because Chloe is smarter. It's because Chloe's earliest dollars spent more years compounding.

This is the single most empowering investing message for a teenager: small, steady contributions at a young age aren't "cute." They're powerful.

The Emotional Catch: Early Investing Feels Like Nothing Is Happening

Here's where most people quit—quietly, politely, without drama.

The first few years of investing can feel underwhelming. If your child invests $25 a week and the account grows to $800, it doesn't look like a revolution. It looks like "money that could have been shoes." The growth feels small because the balance is small. Compounding is shy at the beginning.

But compounding has a personality change later. The growth stops being a small extra and starts being a second engine. In the later years, the account can grow by more in a year than you could contribute with a normal budget. That's the moment people describe as "it took off." It didn't take off. It just finally got big enough for the math to be obvious.

A helpful line to give your child:

> Early investing is boring on purpose. Boring is what building looks like.

A Concrete Case Study: The 401(k) Match That Becomes a Fortune

If you want a real-world example that doesn't require picking stocks or telling scary stories, talk about the most common "free money" many adults ignore: the 401(k) match.

Imagine a 22-year-old starting a first job. The employer offers to match part of what the employee contributes to a retirement plan. Many young workers skip it because rent is real and retirement feels fictional.

Now zoom out. If that worker contributes enough to get the full match, two things happen:

- Their own contributions get decades to compound.
- The match compounds too.

The match isn't just a one-time bonus. It's a seed that gets planted early. Over a 40-year career, an early match can turn into a shockingly large slice of a retirement account—even if the person never becomes a high earner.

This case study lands well with teens because it reframes "benefits" from boring paperwork into a real-life money lever: *some jobs come with an investing head start.*

Compound Growth Has an Evil Twin: Fees

Kids understand "leaks" better than "percentages." So describe fees as a leak in the bucket.

Investment fees often show up as a percentage per year (an *expense ratio* in funds). A fee of 1% doesn't look scary—until you remember two things:

- It's taken *every year.*
- It compounds in reverse by shrinking the amount that gets to keep compounding.

A 1% fee isn't a one-time $1 out of $100. It's a permanent reduction in your account's growth potential. Over decades, that can mean the difference between "retire comfortably" and "work longer than you wanted."

A parent-friendly way to say it:

> Fees are the opposite of compound growth. They're compound drag.

You don't need to turn your child into a fee detective overnight. You just want them to internalize a rule that protects them later when shiny products show up: *if two investments are trying to do the same job, the cheaper one gets a serious head start.*

The Habit That Makes Compounding Real: Automatic Contributions

The easiest way to break compound growth is to rely on willpower. Willpower is not a financial system. It's a mood.

Automatic contributions—money transferred on a schedule—do two quiet, life-changing things:

- They keep you investing through good markets and bad ones, without needing a pep talk.
- They turn investing into a normal expense, like a phone bill, rather than a heroic decision.

For teens, this can be as simple as splitting pay from a part-time job:

> "Every paycheck, $20 goes to long-term investing before I decide what to spend."

Not because $20 is magic, but because the habit is.

A Better Way to Talk About Returns With Kids

Parents sometimes worry that showing growth examples will make kids expect guaranteed results. That's a good instinct. Markets don't hand out promises.

Instead of focusing on a specific return number, focus on the controllable pieces:

- *How early you start*
- *How consistently you add*
- *How often you interrupt the plan*
- *How much you pay in fees and taxes*

Those are levers. They are behaviors. And they're exactly what kids can practice now, even with small dollars.

You can even make it a family game: pick two starting ages and run a quick calculator together. Keep the point simple and emotional: the earlier start doesn't "win" by being better. It wins by being earlier.

Actionable Takeaway: The ``$100 Challenge" That Teaches Compounding Fast

Do this with your kid (or for yourself) sometime this week:

1. Pick a realistic monthly amount: $25, $50, or $100.
2. Choose two starting ages: one early (like 15 or 18) and one later (like 28 or 35).

3. Use the Rule of 72 to estimate how many doublings could happen by age 65 at a few different return assumptions (try 5%, 7%, and 9%).
4. Write one sentence on a sticky note and put it where paychecks get deposited:
   "I can't control the market. I can control my start date."

Compound growth isn't powered by genius. It's powered by time, reinvestment, and the stubborn willingness to keep the loop running—even when the early results look too small to brag about. That's how ordinary amounts of money turn into serious future options.

7.3 Don't Put All Eggs in One Basket

A teen gets their first paycheck and immediately wants to buy something with a story. Not "the stock market." A *stock*. The one their friends talk about. The one that makes headlines. The one that feels like it has a personality.

That impulse is completely human. Stories are sticky. "I own a piece of *that company*" feels cooler than "I own a tiny slice of thousands of companies I can't name." But investing doesn't pay extra for cool. It pays for owning productive businesses *without* letting any single business be able to ruin you.

That's what diversification is: a way to build a life where one bad break doesn't become a financial catastrophe.

The Difference Between Investing and Guessing

Some families don't mind volatility—they've made peace with a bumpy ride. What they can't tolerate is the feeling of being wrong. Picking individual stocks often turns investing into a quiz you have to keep retaking:

> "Did I choose the right company? Did I buy at the right time? Should I sell now?"

That's not a moral failure. It's just a fragile system. The more your outcome depends on being "right" about a specific company, the more your future is tied to surprises you can't control: a new competitor, a lawsuit, a regulation change, a product flop, a CEO scandal, a technological shift. Even a great business can be a terrible investment if the price you paid assumed perfection.

Diversification is the opposite vibe. It says: *I don't need to be right about one thing. I just need the overall economy to keep doing what economies tend to do over long stretches: grow, innovate, and adapt.*

Concentration Risk: The One Problem Diversification Is Built to Solve

There's a specific kind of danger that shows up again and again in real people's finances, and it's not "the market went down." It's concentration.

Concentration risk means too much of your money depends on one bet: one stock, one industry, one country, one real estate market, one employer. When that one bet goes wrong, the damage isn't a temporary dip—it can be permanent.

A kid-friendly way to put it:

> "If one thing can break your whole plan, your plan is breakable."

Diversification makes your plan harder to break.

A Real-World Case Study: When ``Safe" Wasn't Safe at All

Plenty of hardworking adults have made the same mistake with different labels: they loaded up on the company they worked for. It felt loyal. It felt informed. It felt safe—because they saw the brand every day.

And then the company hit trouble.

Sometimes the stock falls because the business declines. Sometimes it falls because the business is fine but the price had gotten too optimistic. Either way, the painful part of this story is the double-hit: when a company struggles, employees can face layoffs or pay freezes at the same time their retirement savings are dropping. Income and investments get hurt together.

That's concentration risk in its most personal form: your job and your portfolio holding hands on the way down.

You don't have to name a specific company for the lesson to land. The point is structural: if your paycheck already depends on your employer, your investments don't need to depend on it too.

Diversification: Not ``Own a Lot of Stuff,'' But ``Own Different Stuff''

People sometimes think diversification means sprinkling money into a bunch of random things—five stocks instead of one, or a handful of trendy names across apps.

That can still be concentrated if those "different" things all move together. Owning five technology companies isn't much diversification if the entire tech sector gets hammered at the same time. Owning a bunch of meme coins isn't diversification; it's just chaos with extra steps.

Diversification works when your investments are meaningfully different from each other—different companies, different industries, sometimes even different types of investments—so that a single blow doesn't land everywhere at once.

This is where you can teach a powerful concept without using fancy words: *not everything reacts to the same news the same way.* Gas prices can rise while streaming subscriptions get canceled. Health care demand can be steady while luxury spending drops. Different businesses feel the economy differently.

Index Funds: ``Buying the Market'' Instead of Playing Stock Scout

An index fund is one of the most parent-friendly inventions in modern finance. It's a fund designed to track a market index—basically a scoreboard that represents a large chunk of the market (like the S&P 500, which includes hundreds of large U.S. companies). Instead of trying to pick winners, an index fund tries to *own the whole game.*

If you buy a broad U.S. stock index fund, you're not betting on one company's product launch. You're buying a small piece of many companies across many industries. If one company stumbles, it's a speed bump, not a cliff.

That's why index funds are often described as "boring." And in family investing, boring is not an insult. Boring is a safety feature.

Why Index Funds Can Feel Too Simple (and Why That's the Point)

There's a weird insecurity that shows up around money: if something is simple, we assume it can't be powerful. We equate complexity with sophistication. Kids pick this up from culture—apps with confetti animations, influencers talking about "strategies," headlines that make investing sound like a high-speed sport.

Index funds are like a seatbelt. Nobody brags about wearing one. But it's one of the smartest things you can do every time you get in the car.

Owning the market is not about being lazy. It's about refusing to gamble with the parts of your future that don't need gambling.

Diversification Doesn't Mean ``Nothing Bad Can Happen"

It's tempting to oversell diversification as a shield against losses. That would be comforting—and misleading.

Diversification can't prevent a broad market from dropping. In a serious recession, many stocks can fall together. Index funds will go down too. Diversification reduces the chance that a single failure wrecks you, but it does not erase volatility.

That distinction matters for kids because it teaches them to expect normal turbulence. A diversified investment can still have ugly stretches. The goal isn't to avoid every downturn. The goal is to avoid the kind of risk that permanently knocks you out of the game.

The ``One-Basket" Trap Kids Fall Into: Familiarity

Ask a teenager which company they'd buy, and they'll usually pick something they personally use: their favorite sneakers, their phone brand, the place they order food, a gaming company. Familiarity feels like knowledge.

But knowing a product is not the same as knowing a stock. The stock price reflects not just whether the company is good, but whether the company does *better than everyone already expects*. A company can be beloved and still be overpriced. A company can be boring and still be a great investment.

Index funds help kids sidestep this trap. They can still talk about companies and learn what businesses do—but their financial outcome doesn't depend on guessing which brand will be the hero this decade.

Asset Allocation: The Mix Matters Too

Diversification is about spreading within a category (like stocks across many companies). Asset allocation is about mixing categories (like stocks and bonds, or stocks and cash). You don't need a finance degree to understand the point: different types of investments behave differently.

Stocks tend to be bumpier but have historically had higher long-term growth potential. Bonds tend to be steadier but usually grow more slowly. Cash is stable in dollar terms but vulnerable to inflation over long periods.

For a teen, the best takeaway is practical: *the closer you are to needing the money, the less you want it riding a roller coaster.* For long-term "Later Money," a higher stock allocation can make sense; for near-term goals, stability matters more than growth.

Correlation: Why ``Different" Sometimes Isn't Different Enough

Here's a sneaky truth: in scary moments, many investments start acting like they're all the same. That's because fear is contagious.

In calm times, one industry can be up while another is down. In crises, investors sell lots of risky assets at once. This is why diversification is a reduction of risk, not an on/off switch.

But correlation—the tendency of investments to move together—still matters. It's why "I own 12 stocks" isn't automatically safer than "I own an index fund." If those 12 stocks are all in one sector, or all sensitive to the same economic forces, you may not be as diversified as you think.

The simple family rule:

> "Count *types* of exposure, not just the number of tickers."

Rebalancing: The Boring Maintenance That Keeps Risk From Sneaking Up on You

Even a well-diversified portfolio can drift. If stocks have a great year, they can become a larger portion of your portfolio than you intended. That means your risk level quietly increased—without you making a conscious decision.

Rebalancing is the act of bringing your portfolio back to its target mix. Some people do it on a schedule (once a year). Others do it when the mix moves beyond a certain threshold. Either way, rebalancing is a discipline, not a prediction. You're not saying, "Stocks will fall." You're saying, "I don't want my portfolio to become riskier just because stocks had a good run."

For kids, this is an unusually good lesson in adulthood: maintenance beats drama. The strongest financial plans are more like brushing teeth than like winning a contest.

Costs Still Matter: Diversification Doesn't Fix High Fees

It's possible to buy something diversified and still get a bad deal. If a fund charges high fees year after year, those fees quietly drain returns. Compounding works both ways: it grows your gains, and it grows the impact of what you lose.

A diversified, low-cost index fund can be a double win: broad exposure and less "drag." Meanwhile, a diversified fund with high expenses can still leave you behind—even if it sounds impressive in marketing language.

A practical habit to teach:

> "Before buying a fund, look for the expense ratio and ask, 'What am I paying every year for this?'"

A Concrete Example: The ``Teen Stock Pick" vs. The ``Market Slice"

Imagine two siblings get $500 each from a grandparent.

Sibling A buys $500 of a single company—let's say a hot name everyone recognizes. Sibling B buys $500 of a broad market index fund.

A year later, the single stock is down 40% because the company missed earnings and investors got spooked. The index fund is down 8% because the overall market had a rough year too.

Sibling A is now at $300. Sibling B is at $460.

Both took risk. Both faced volatility. But the *kind* of risk was different. Sibling A took company-specific risk—one storyline, one set of surprises. Sibling B took market risk—many storylines at once. Over long periods, market risk is the one investors are typically paid to take. Company-specific risk is the kind you often take for free.

Now flip it. Sometimes the single stock will soar and beat the index fund. That's the hook. It's also the trap: the occasional big win makes the strategy feel "smart," even if it's unreliable. In a family plan, reliability matters. A kid doesn't need to win a lottery ticket. They need a system that works even when they're busy, tired, or not thinking about money at all.

How to Talk About Index Funds Without Making Them Sound Like Vegetables

If you pitch index funds like broccoli—good for you, eat it because I said so—your kid will treat investing like a chore.

A better framing is to make it about *freedom*.

Index funds are a way to own a slice of human progress without having to predict which brand will dominate it. You're investing in the idea that people will keep solving problems: building better tools, curing diseases, improving logistics, inventing products we can't imagine yet. Some companies will fail. New ones will rise. An index updates over time, which means your basket evolves as the economy evolves.

That's a surprisingly hopeful thing to teach a teenager in a world full of noise: you don't have to pick the future perfectly to participate in it.

A Parent-Child Practice: The ``Eggs and Baskets" Conversation That Sticks

Try a short exercise that turns diversification into something your child can feel.

Ask: "If you could only invest in one thing, what would it be?" Let them pick. Then ask: "What would have to go wrong for that to be a bad investment?" Keep going until they realize how many ways the future can surprise a single company.

Then ask: "What if you invested in *a lot* of companies at once?" The answer becomes less dramatic: "Some would do badly, but not all."

You're not trying to scare them away from individual stocks forever. You're teaching them to recognize the trade:

- Picking stocks: more exciting, more concentrated risk, more dependence on being right.
- Index funds: less exciting, more diversified, more dependent on time and patience.

Excitement is not a retirement plan.

Actionable Takeaway: A Simple Rule for Real Money

Here's a rule that fits on an index card and can guide a family for decades:

> *Make your "default investment" a low-cost, broad index fund. If you want to pick individual stocks for learning, keep it small enough that a mistake won't change your life.*

Then make it concrete. Choose a percentage that matches your household—something like 90/10, where 90% goes to the diversified "basket" and up to 10% can be "curiosity money" for individual stock picks. Curiosity money is where kids can practice without the practice turning into damage.

When your child learns to spread risk on purpose—to build a basket that can take a hit and keep going—they're not just learning investing. They're learning a life skill: how to protect their future from the fact that the world is surprising.

7.4 Accounts for Growth

If investing is the engine, an account is the vehicle you bolt it into. Same engine, wildly different driving experience.

That sounds abstract until you're a parent holding a check from Grandma made out to your eight-year-old, or until your teenager comes home from a first job with a paystub and a grin that says, "I did a real thing." Suddenly the question isn't "*Where* do we put this so it grows without creating a mess later?"

The most useful way to think about accounts is as *wrappers*. The investments inside the wrapper might be the same (often a simple index fund), but the wrapper changes the rules: who controls the money, when it can be used, and how much of the growth you get to keep after taxes.

Two wrappers come up constantly for families who want to give kids a head start: custodial accounts and Roth IRAs for kids. They sound like paperwork. They are, secretly, parenting tools.

The Parenting Problem These Accounts Solve

Kids don't naturally think in decades. They think in weekends. Even responsible kids can struggle to connect "I saved $200" with "I will be grateful for this in twenty years."

Accounts help you turn long-term thinking into something more concrete. They create separation between:

- money a child can spend soon, and
- money meant to become a future advantage.

That separation matters because one of the biggest threats to compounding isn't the market. It's temptation. A good account structure keeps "future money" from being constantly renegotiated every time a new want shows up.

Custodial Accounts (UGMA/UTMA): The Flexible, Slightly Scary Gift

A custodial account is a brokerage or bank account opened for a minor, managed by an adult custodian (usually a parent) until the child reaches the age of majority in your state. You'll often see the acronyms UGMA or UTMA-legal frameworks that allow minors to own assets with an adult managing them.

Here's the plain-English version:

> A custodial account lets you invest in your kid's name now, while you steer the ship-until the day the ship legally becomes theirs.

That last clause is the whole story. Custodial accounts are powerful because they're simple and flexible. They're also a little terrifying because they come with a built-in handoff date.

Why families love custodial accounts

They're flexible. Money in a custodial account can generally be used for the child's benefit, not just education. That means it can help with summer programs, a laptop, a car for commuting, a gap-year plan, or a first apartment deposit-depending on your judgment and the rules you follow as custodian.

They're real ownership. The account belongs to the child. That can be motivating for kids who light up when something is truly "theirs." It also makes for great financial education: statements, contributions, gains, taxes-real life, scaled down.

They allow investing early. If your goal is simply to get a child invested in the market with small amounts, custodial accounts are often the most direct route.

The tradeoff families underestimate

At the age of majority (often 18 or 21, depending on the state and account type), control transfers. Not "in theory." Not "if they're mature." It transfers.

That means the account you carefully grew for a decade could become, in the wrong moment, the down payment on a used sports car with expensive insurance and a mysterious smell. Most kids won't do that. But the fact that they *can* changes how you should think about the size of the account and how you teach responsibility along the way.

A good mental model:

> A custodial account is a gift with a timer.

Timers are not bad. They just require planning.

Roth IRA for Kids: The ``Head Start'' Account With a Catch

A Roth IRA is a retirement account where contributions are made with money that has already been taxed, and qualified withdrawals in retirement can be tax-free. The appeal is obvious: if you can get money into a Roth early, decades of growth may escape future taxation.

So why doesn't every parent open a Roth IRA for every kid?

Because of the catch: the child must have *earned income*.

Earned income means money from working-wages from a job, self-employment income from legitimate work (like babysitting, tutoring, mowing lawns), and similar. It does *not* include allowance, gifts, or investment income. Your teen can't get a $1,000 birthday check and call it "earned." The IRS has a very particular sense of humor about that.

The easiest way to explain this to a kid:

> A Roth IRA is for money you earned by doing something, not money you got because someone loves you.

If your child has earned income, a custodial Roth IRA (an IRA opened with an adult as custodian) can be one of the most valuable accounts you can help them start.

The two superpowers of a Roth IRA for a young person

1) *Time turns it into a monster (in a good way).* A small contribution at 16 has far more years to compound than a larger contribution at 35. This is the compounding story in its purest form: tiny now, enormous later.

2) *Contributions (not the growth) are generally accessible.* In many cases, Roth IRA contributions can be withdrawn without taxes and penalties because they were contributed after tax. (The

growth has stricter rules.) This feature is often misunderstood, and it should never be an excuse to treat a Roth like a checking account-but it can reduce the fear that money is "locked away forever."

Still, the best use of a Roth IRA is exactly what it says on the label: retirement. The point of opening one for a teen isn't to create a flexible spending pool. It's to give them a financial advantage most adults wish they'd started earlier.

A Concrete Case Study: Mia's Summer Job and the $1,200 Question

Mia is 16 and works at a local pool snack bar during the summer. Over three months she earns $2,400. She wants to spend it-understandably-on a used laptop and going out with friends. Her parents want her to learn investing without turning money into misery.

They make a deal.

Mia will put $100 a month into a custodial Roth IRA during her working months, for a total of $300. Her parents will "match" it with $300 as a reward for saving-*but the contribution can't exceed her earned income.* The Roth IRA ends the summer with $600 invested.

Mia still buys a laptop. She still lives her life. But now she owns something unusual for a 16-year-old: a retirement account.

Here's what matters about this case study:

- The contribution was tied to real work, which makes it feel earned rather than imposed.

- The amount was small enough to be doable, large enough to be meaningful.
- The match turned saving into a win, not a punishment.

Most importantly, Mia now has a mental category called "money that grows while I sleep," and it's connected to a memory: that summer she worked, and part of her paycheck went to future freedom.

Custodial Brokerage vs. Custodial Roth IRA: The Cleanest Comparison

Parents often get stuck because both options sound good. Here's the simplest way to separate them:

A custodial brokerage account is for the child's future in general. Flexible, usable for many purposes, taxable in the normal way for investment accounts.

A custodial Roth IRA is for the child's retirement future specifically. Potential tax advantages are enormous, but contributions require earned income and the account is meant to be left alone for a long time.

If your child has earned income and you're choosing where their "investing habit" should live, a Roth IRA often deserves first consideration because of the tax-free growth potential over decades. If your child does *not* have earned income, the Roth IRA is simply not an option for contributions, and a custodial brokerage account becomes the practical default for long-term investing in the child's name.

A Word About 529 Plans (Because Someone Will Ask)

Many families also use 529 education savings plans, which can offer tax advantages when used for qualified education expenses. They can be great tools-especially when college is a likely goal and you want guardrails around education spending.

But 529s are not the same as "accounts for investing." They are education-focused, with their own rules and tradeoffs. A custodial brokerage account and a Roth IRA are broader investing frameworks; a 529 is a targeted tool.

If you already have a 529 humming along, that doesn't make custodial accounts or a Roth IRA pointless. It just means you're building different kinds of future options.

The ``Earned Income" Conversation: How to Do It Without Making It Weird

The moment you mention a Roth IRA for a kid, parents ask: "Can I pay my child for chores?" Sometimes they mean it innocently. Sometimes they're hoping for a clever workaround.

The healthiest approach is also the simplest: keep it honest.

Regular household chores (cleaning their room, doing dishes) are typically part of being in a family, not a job. But paying a child for legitimate work can be reasonable: babysitting, tutoring, dog walking, mowing, helping with a family business in a real, trackable way.

What matters is that it's real work, the pay is reasonable for the work performed, and it's documented appropriately. If you feel tempted to create "earned income" out of thin air, that's a sign to slow down and do fewer clever things with your taxes and more consistent things with your habits.

Kids learn a deeper lesson here:

> If you want the special account, you have to do the real work.

That's not just tax compliance. That's character training.

Control, Values, and the Moment the Account Becomes Theirs

If you choose a custodial account, you are choosing a future handoff. Treat that handoff like you would treat teaching a teenager to drive. You don't wait until the day they get the keys to mention brakes.

Start early with shared visibility:

- Show them the statement occasionally.
- Let them see the investments (ideally simple, diversified funds).
- Talk about what the money is *for*.

As they get older, give them partial responsibility:

- Let them decide what portion goes to spending, saving, and investing.
- Let them make a small investing choice with guardrails (for example, choosing between two broad index funds rather than between random stocks).

The goal is not to raise a child who never makes a dumb purchase. The goal is to raise an adult who can hold the keys without immediately crashing into a financing trap.

Taxes and Paperwork: The Part Nobody Posts About

It's not glamorous, but it's worth saying out loud: custodial brokerage accounts can generate tax forms. Even if the amounts are small, dividends and capital gains can trigger reporting. Many families are surprised the first time their child's investing account produces a tax document.

This isn't a reason to avoid investing. It's a reason to keep the portfolio simple, keep good records, and be ready for a little extra admin. If you want "as easy as possible," a Roth IRA (when available) can be simpler from a taxes-on-growth perspective, while still requiring normal account reporting and good habits.

The best emotional framing for parents:

> Paperwork is not a problem. It's proof we're doing real life.

How Much Should Go Into a Kid's Account? The Answer That Prevents Regret

Parents sometimes swing between extremes:

- "We should put everything we can into our kid's account."
- "What if they waste it? Maybe we shouldn't put much at all."

A more balanced approach is to decide what you're trying to buy with the money.

If you want to buy a financial *head start*, you don't need a massive account. A modest Roth IRA started young can do astonishing work on its own.

If you want to buy *flexibility*-help with a first car, a security deposit, a move for an internship-a custodial brokerage account can serve that role, as long as you accept the control transfer.

If you want to buy *education*, a 529 may be the more targeted tool.

You don't need to solve every future need with one account. You need to choose one account that fits your next dollar and your family's values.

A Simple Decision Rule That Works in Real Kitchens

When the question is "Which account should we open for our kid?" use this rule:

> *If the child has earned income, consider a custodial Roth IRA first for long-term investing. If the child does not have earned income, a custodial brokerage account is the straightforward way to invest in the child's name.*

Then add the parent part-the part no account can do for you:

> *Keep it simple, keep it diversified, and teach the child what the money is for long before the control transfer happens.*

Actionable Takeaway: Set Up One ``Future Money" Account and One Habit

Choose one account you can actually maintain without resentment-either:

- a custodial brokerage account if your child doesn't have earned income, or
- a custodial Roth IRA if your child does have earned income.

Then set one habit that makes the account come alive:

> *Every time money comes in (gift, paycheck, or allowance), a small, pre-decided slice goes to Future Money before anything gets spent.*

Start with a number that won't cause a fight-$10 a week, $25 a month, whatever is realistic-and automate it if you can. The habit is the real inheritance. The account is just the container that keeps the habit growing until it's strong enough to stand on its own.

Chapter 8

The Launchpad: Real-World Independence

Welcome to the real world, where the first paycheck is always smaller than expected and the landlord wants three months of rent upfront. This chapter is the 'missing manual' for adulthood. We move past theory into the gritty logistics of independence: deciphering tax withholdings, signing legally binding leases, understanding why cars are wealth-destroying machines, and building a safety net that catches you when you inevitably trip. It is time to clear the runway for take-off.

8.1 Decoding the Paycheck

The first paycheck has a special talent: it can make a smart young adult feel temporarily betrayed by math.

They remember the hourly rate they were promised, or the salary on the offer letter. They did the mental shopping-cart math. Then payday arrives and the deposit is smaller—sometimes *way* smaller. The gap can feel like a mistake, or worse, a trick. It's neither. It's a system: money flows through a set of gates, and each gate has a name, a rulebook, and a reason it exists.

Once your kid understands those gates, a paycheck stops being mysterious. It becomes a tool they can read, predict, and audit—which is exactly the skill adulthood demands.

Gross pay vs. net pay (the only two numbers that really matter at first)

Gross pay is the headline number: what they earned before anything is taken out. It's the pay rate times hours (for hourly work), or salary divided into pay periods (for salaried work). Gross pay is what your kid *generated*.

Net pay is what actually lands in their account. It's what they can spend without overdrafting. Net pay is what your kid can *live on*.

Everything in between is the "missing money," and it isn't random. It falls into two big buckets:

- *Taxes withheld* (money sent to the government on their behalf).
- *Deductions* (money routed to benefits or savings plans they chose, or that are required).

If your kid learns only one habit here, make it this: *budget off net pay, plan off gross pay.* Net tells you your monthly reality. Gross tells you your true earning power and what you can change over time.

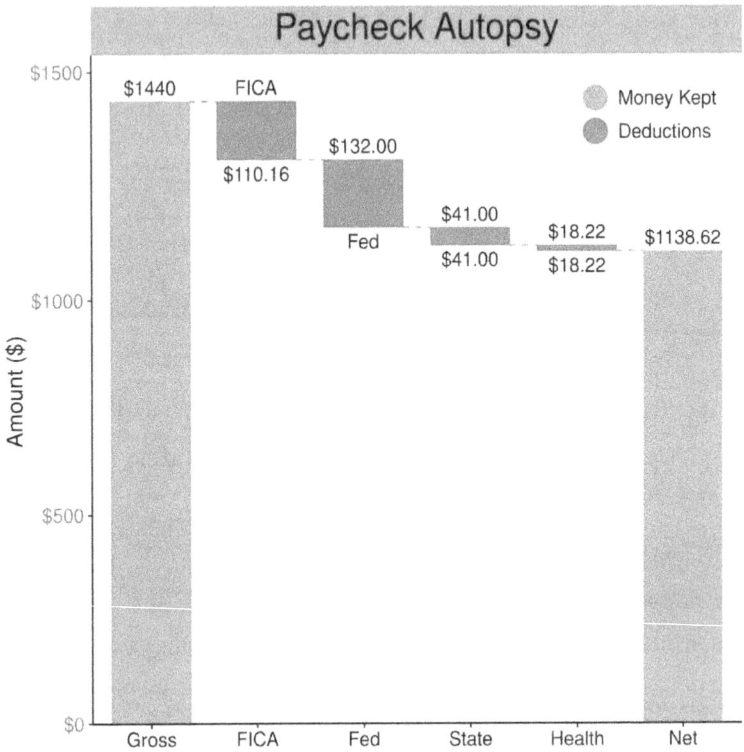

That picture is the mental model. Now make it real.

A concrete case study: Maya's first "real" pay stub

Maya is 19. She works 40 hours a week at $18/hour at a sports medicine clinic. She gets paid every two weeks.

Her gross pay should be:

$$80 \text{ hours} \times \$18 = \$1{,}440$$

On payday, she sees $1,138.62 hit her checking account. Maya's first thought is that payroll messed up. Her second thought is that adulthood is a scam. The pay stub clears it up: nothing broke; several things happened *on purpose*.

Below is a simplified version of how a typical stub reads. (Real stubs vary, but the categories rhyme.)

Line item	*This paycheck*	*What it is*
Gross earnings	$1,440.00	Pay before withholdings/deductions
Social Security tax	-$89.28	6.2% of wages (employee share)
Medicare tax	-$20.88	1.45% of wages (employee share)
Federal income tax (withheld)	-$132.00	Prepayment toward annual federal tax
State income tax (withheld)	-$41.00	Prepayment toward annual state tax
Health insurance premium	-$18.22	Employee cost for health plan (if enrolled)
401(k) contribution	-$0.00	Retirement savings (if elected)
Net pay	*$1,138.62*	Amount deposited

Look at what this does to Maya's life planning. If she's apartment-hunting or trying to decide whether she can handle a car payment, the only number that matters is net pay. Her budget can't spend gross. Her bills don't accept "but my offer letter said..."

The goal isn't to make her feel smaller. It's to give her a reliable dashboard.

FICA: the deduction with the least drama and the most certainty

FICA sounds like a mysterious creature from the tax swamp, but it's just two taxes with predictable rules:

- *Social Security tax:* 6.2% of wages (up to a wage limit that changes over time).
- *Medicare tax:* 1.45% of wages (no wage limit).

On the stub, these usually show up as separate lines (Social Security and Medicare). Together, many people casually call them "FICA." Your kid doesn't need to memorize the acronym to understand the point: this isn't negotiable, and it's not based on their "tax bracket." It's a flat percentage (for most earners) that comes out early, like a toll road.

Here's the parenting magic trick: ask your kid to estimate FICA in their head.

If they earn $500 gross in a week, FICA is roughly:

$$\$500 \times 7.65\% \approx \$38$$

That one mental estimate prevents the classic mistake of spending as if the whole $500 is available. It also turns "paycheck shock" into "oh, that's the toll."

Also: employers match these taxes on their side. Your kid won't see that money on the stub because it's not paid *from* their wages, but knowing it exists helps them understand why hiring isn't just "wages times hours" for a business.

Federal and state withholding: not your tax bill, just your tax prepayment

The most misunderstood line on a pay stub is federal income tax. Many young workers see "Federal Withholding" and assume: *This is my tax. This is what I owe.*

It's not. It's a *deposit.*

Income tax in the U.S. is calculated across the whole year. Payroll can't see your whole year in January. So it withholds a best-guess amount each paycheck based on the information your kid provides on Form W-4 (filing status, dependents, extra withholding, and whether they have multiple jobs).

That's why two people with the same wage can have different take-home pay:

- One checks a box that tells payroll to withhold more (bigger refund later, smaller paycheck now).
- Another claims credits or has a second job and adjusts the form (different withholding pattern).
- One lives in a state with income tax; another doesn't.

This is where you can give your kid a surprisingly grown-up advantage: teach them to treat withholding like a thermostat, not a mystery.

Too little withheld can lead to a nasty surprise at tax time: they might owe money in April, sometimes with penalties if they really under-withheld.

Too much withheld can lead to a big refund—which feels great for about ten minutes until you realize what it means: the government held their money all year. A refund isn't "free money." It's change from an overpayment.

A healthy, boring goal for many young workers is a small refund or small amount owed. Not because either is morally superior, but because it means their paycheck matches reality.

The W-4: the quiet form that decides your take-home pay

If your kid wants one "adult lever" to pull early in life, it's the W-4. The W-4 doesn't decide their tax rate. It decides how much is withheld *during* the year.

If they start a job and don't submit a proper W-4, payroll often defaults to a standard setting that can be unhelpful (sometimes too much, sometimes too little). That's why the first few checks can look strange compared to later ones.

A practical family habit: the first time your kid gets paid at a new job, sit down with the pay stub and ask two questions:

1. "Does the gross pay match your hours and pay rate?"
2. "Does the withholding pattern make sense for your situation (one job vs. multiple, dependents, etc.)?"

This is not about micromanaging. It's about catching errors while they're small. A wrong hourly rate for two weeks is annoying. A wrong hourly rate for six months is a slow financial leak.

Pre-tax vs. post-tax deductions: the two flavors of 'missing money'

Taxes are one kind of "gone." Benefits are another. Benefits deductions can be either *pre-tax* or *post-tax*, and that tiny prefix changes the math.

Pre-tax means the money is taken out before certain taxes are calculated. For example, some health insurance premiums and certain retirement contributions reduce "taxable wages," which can lower the federal/state income tax withheld (and sometimes other taxes depending on the benefit type). Pre-tax can be a discount, because it shrinks the portion of income that gets taxed.

Post-tax means the money comes out after taxes. It's still a purposeful route, but it doesn't reduce taxable income.

Your kid doesn't need a tax-law seminar. They need a common-sense way to read the stub:

- Taxes = required payments based on earnings.
- Benefits = chosen (or employment-based) costs and savings routes.
- Pre-tax benefits = often more efficient than paying with after-tax dollars.

Retirement contributions: the weirdly powerful deduction that looks like a loss

A 401(k) contribution is the rare deduction that is both "money leaving the paycheck" and "money you still own." It doesn't vanish; it changes containers.

This is a hard emotional sell to a young adult because it feels like taking a bite out of today for the sake of a foggy future. The way to make it real is to frame it as buying future freedom in small, painless chunks.

If your kid contributes $50 per paycheck, they probably won't remember the $50 in a month. But they will remember the moment, years later, when they check an account and realize: *I built something.*

Also, if an employer offers a match, it's not a cute perk. It's part of compensation. Turning down a match is like refusing a portion of your paycheck because the paperwork is annoying.

Overtime, bonuses, and the myth of the "higher-tax paycheck"

Young workers often hear this sentence from a coworker who sounds confident:

> "If you work overtime, you'll get taxed more and it won't be worth it."

Here's the truth that keeps people from making career decisions based on breakroom folklore:

Working more can increase withholding on that particular paycheck, but it doesn't usually make the extra work pointless.

Withholding is an estimate. When a paycheck is unusually large (overtime week, bonus, commission spike), the payroll system may withhold as if that large check is your "normal" all year. That can make the net feel disappointing. Later, when you file a tax return, the annual math is done using your *total* income. If too much was withheld during the spike, it often comes back as part of the refund.

So if your kid works an overtime-heavy month and says, "They took so much more—it's not worth it," you can answer with a calm adult sentence:

"That's withholding. The real math happens on the tax return."

Why filing a tax return matters even when your kid is "just working part-time"

Many young adults think tax returns are only for "real adults," homeowners, or people with complicated finances. But filing can matter even with simple jobs for three big reasons.

1) Refunds: getting back what was over-withheld

If your kid had federal income tax withheld but earned little enough that their actual tax bill is low (or zero), they may be owed a refund. The government doesn't automatically mail it just because they were nice enough to overpay. They have to file to claim it.

This is the simplest reason to file: it's the only way to reconcile what was withheld with what was actually owed.

2) Proof and paperwork: a tax return is a financial passport

Tax returns are boring until the moment you need one. Landlords, scholarship programs, financial aid offices, and lenders often ask for tax documents because they're standardized proof of income.

A young adult who files on time is quietly building a file cabinet of credibility.

3) Catching problems early: identity and reporting errors don't fix themselves

Sometimes a W-2 is wrong. Sometimes an employer reports something incorrectly. Sometimes a young adult gets a tax form from a job they swear they never had (a potential red flag for identity theft). Filing is how those issues surface while they're still fixable.

Think of a tax return like reconciling a bank account: it's not glamorous, but it's how you notice the leak.

A quick tour of the forms your kid will hear about (without turning it into a class)

Most first-job tax life boils down to a few documents:

- *Pay stub:* the per-paycheck receipt (gross, taxes, deductions, net).

- *W-2:* the annual summary from an employer (usually arrives in January for the prior year).
- *1099 forms:* common for gig work/contracting, where taxes often *aren't* withheld automatically.

A useful parent move is to connect the dots: pay stubs stack up to become a W-2. The W-2 feeds the tax return. The tax return is where the year gets "graded" and any overpayment comes back.

How to teach a kid to read a pay stub like a grown-up

If your kid can read a nutrition label, they can read a pay stub. It's the same skill: scan for the handful of lines that matter most and ignore the rest until needed.

Have them highlight these items:

1. *Pay rate and hours:* Are they correct?
2. *Gross pay:* Does it match what they expected from hours × rate?
3. *Net pay:* Does it match what hit the bank?
4. *FICA lines:* Social Security and Medicare (predictable percentages).
5. *Federal and state withholding:* The adjustable prepayment.
6. *Benefit deductions:* Health, retirement, HSA/FSA, etc.

Then ask one question that changes behavior fast:

> "Which of these numbers would you use to decide what you can afford *every month*?"

When they answer "net pay," you've just saved them from the most common early-adulthood budgeting mistake.

Actionable takeaway: the 10-minute paycheck checkup

On your kid's next payday, do a 10-minute checkup together:

1. Pull up the pay stub and the bank deposit amount; confirm net pay matches the deposit.
2. Recalculate gross pay from hours × rate (or salary/pay periods) to confirm nothing is off.
3. Estimate FICA as gross × 7.65% and see if the stub is in the ballpark.
4. If withholding seems oddly high or low, have them review their W-4 info (especially if they have more than one job).
5. Write down one number on a sticky note (or in a notes app): *net pay per paycheck*. That becomes their real budgeting baseline.

A paycheck isn't just money arriving. It's a report card on a system. Once your kid can read it, they stop guessing—and guessing is where most "I don't know where my money went" stories begin.

8.2 Moving Out: Renting and Leases

The first apartment is supposed to feel like freedom. In practice, it often feels like a bouncer at a club you're not sure you can afford: *"Name on the list? Credit check? Deposit? First month?*

Possibly last month? Also we need a $75 application fee and proof you make three times the rent."

Young adults tend to think renting is a simple monthly bill. Landlords and property managers think renting is a risk-management business. That mismatch is where a lot of panic, shame, and bad decisions get born—especially the kind of bad decision that starts with "I didn't want to ask my parents again."

The best way to make renting feel fair is to treat it like what it is: an upfront cash event, a credit event, and a legal event. If your kid can see those three events coming, they can plan instead of scramble.

The ``entry fee" problem: rent is not the price of moving in

Ask a young adult what an apartment costs and they'll say, "It's $1,600 a month." That's like saying a concert ticket costs $120 and forgetting parking, service fees, and the fact that you also need a ride home.

Most apartments require a pile of money *before* a key ever touches your palm. Common components:

- *Application fee:* Paid to process paperwork and run screening. Often nonrefundable.
- *Holding fee:* Paid to take the unit off the market. Sometimes applied to move-in costs; sometimes not.
- *Security deposit:* Often one month's rent (sometimes more), intended to cover damage beyond normal wear and tear.

- *First month's rent:* Usually due at signing or before move-in.
- *Last month's rent:* Required in some situations, or used as a workaround when credit is thin.
- *Pet deposit / pet rent:* A deposit up front, plus an extra monthly charge.
- *Utility deposits and connection fees:* Especially if the tenant has no prior history with the utility company.

None of this is "gotcha" money. It's the rental market's way of asking, *Can you handle a sudden bill without falling apart?* The uncomfortable part is that many people only discover this question after they've already emotionally moved into the place.

A practical rule of thumb you can hand your kid: *before you tour seriously, assume you'll need* 3×—4× *the monthly rent in cash.* In some cities and situations it's less. In others—especially with weak credit, roommates, pets, or competitive markets—it can be more.

Case study: Jordan's ``affordable" $1,450 apartment that required $4,200

Jordan is 22, starting a job with a steady paycheck. He finds a clean one-bedroom listed at $1,450. It feels responsible. He has $2,300 saved and thinks, *Great, I'm ahead.*

Here's what the leasing office requests:

Move-in item	*Amount*
Application fee	$60
Holding fee (credited to deposit if approved)	$200
Security deposit	$1,450
First month's rent	$1,450
Electric deposit	$250
Renter's insurance first month (paid up front)	$20
Total cash needed before move-in	*$3,430*

That's the "just to get in the door" money. Then comes the stuff nobody puts in the listing:

Hidden-ish setup costs	*Amount*
Basic cleaning supplies, trash can, light bulbs	$80
Shower curtain, towels, bath mat	$90
A cheap toolkit (or repeated emergency trips)	$40
Toilet paper, paper towels, pantry basics	$120
A used table + chair (bare minimum)	$150
A mattress situation (new or used)	$350
Move-in fuel / truck / pizza bribes	$120
Total "life starts now" costs	*$950*

Jordan's "$1,450 apartment" asks for roughly $4,380 in the first couple weeks. His $2,300 savings isn't "ahead." It's not even close. He has three options: ask family, go into debt, or walk away.

This is why moving out feels like it punishes responsible kids. They did the right thing by picking a reasonable rent; they just didn't understand the entry fee. Once you teach the entry fee concept early, saving becomes targeted instead of vague.

Rent-to-income ratios: why landlords keep saying ``$3\times$ rent"

A common screening rule is that gross monthly income should be around three times the monthly rent. It's not a moral judgment. It's a blunt tool: the landlord wants confidence that rent won't be competing with groceries, car repairs, and every other real-world bill.

If rent is $1,500, "3× rent" implies $4,500 of gross monthly income, or $54,000 per year.

Here's the twist: your kid might meet the requirement on paper and still feel squeezed in real life, because landlords usually look at *gross* income. Bills get paid with *net*. If your kid just had their first "gross vs. net" awakening, this becomes an immediate lesson in why take-home pay matters. A place that looks affordable by the landlord's math can still be stressful by the tenant's reality.

A useful conversation isn't "Can you get approved?" It's "If you get approved, will you still be able to breathe?"

The lease isn't a vibe. It's a debt contract.

A lease often gets treated like a permission slip: sign, get keys, hang art, live life. But legally, a lease is closer to a loan contract than a permission slip. It's a promise to pay a fixed amount over a fixed period, with penalties if you don't.

That matters because it reframes the stakes. Breaking a lease can mean:

- Owing rent until the unit is re-rented (depending on state rules and lease language).

- Paying a lease-break fee (sometimes one or two months' rent).
- Losing the security deposit.
- Having an unpaid balance sent to collections, which can damage credit.

Your kid doesn't need to become a lawyer. They need to recognize the core fact: *the lease turns "renting" into a legally enforceable obligation.*

One parenting move that pays off: before they sign, ask them to find and read these clauses out loud:

- *Term:* start date, end date, and what happens at renewal.
- *Total rent due and due date:* plus any grace period and late fee schedule.
- *Security deposit rules:* what counts as damage, what counts as "normal wear," and how return timing works.
- *Early termination / lease break:* fees, notice requirements, and responsibilities.
- *Roommate joint liability:* whether each tenant is responsible for the entire rent.

That last one deserves its own moment.

Roommates: the ``fun" decision with the most expensive fine print

Roommates are often framed as a lifestyle choice. Financially, they're a partnership. And many leases make that partnership *joint and several*—meaning each roommate can be held responsible for the full rent, not just their "share."

Translation: if rent is $2,400 and your kid's roommate disappears, the landlord may still expect the full $2,400 on time. The landlord didn't choose the roommate; your kid did.

This is not an argument against roommates. It's an argument for a roommate agreement that covers:

- How rent is paid (one person collects? each pays portal separately?).
- What happens if someone is late.
- How utilities are split and tracked.
- Guest rules that prevent a "third roommate" from moving in unofficially.
- Move-out timing and cleaning expectations.

The tone matters. Make it casual but explicit: "I like you. I also like not getting sued." Good friendships survive clarity. Fragile friendships require vagueness, which is not a great roommate strategy.

Credit screening: what they're looking for (and what you can do if credit is thin)

Landlords often run a consumer report—typically credit, sometimes rental history, sometimes a background check depending on the property and local rules. Again, it's not personal. It's underwriting.

If your kid's credit history is thin (common for young adults), approval often depends on the rest of the application:

- Strong, stable income.
- Cash reserves.
- A co-signer or guarantor (someone promising to pay if your kid doesn't).
- A higher security deposit (where allowed).

Here's the part that keeps the power balanced: if a landlord takes "adverse action" based on a consumer report—denies the application, requires a co-signer, charges a higher deposit, offers worse terms—they generally must provide an adverse action notice. That notice should tell your kid which consumer reporting company provided the report and how to request a copy.

Why does that matter? Because errors happen. Mixing up two people with similar names happens. Old debts that should have been removed sometimes linger. If your kid gets denied and never asks why, they lose the chance to fix a problem that could haunt them for years.

The most adult move after an application goes sideways is not to spiral. It's to request the notice, get the report, and verify what the landlord saw.

Security deposits: not ``extra rent,'' not a gift, and not a free-for-all

Many young renters treat the deposit like a fee they'll never see again. Landlords sometimes treat it like a blank check for renovations. Reality is somewhere in the middle, and the difference is documentation.

The security deposit is intended to cover damage beyond normal wear and tear and sometimes unpaid rent (depending on local law and lease terms). Your kid can protect it with a simple system:

1. *Do a move-in inspection immediately.* Use the property's checklist if they provide one.
2. *Take timestamped photos and a short video.* Floors, walls, appliances, inside the oven, under sinks, and any existing dings.
3. *Put maintenance requests in writing.* Even if they call, follow up via email or the portal.
4. *Do the same on move-out.* Photos, video, and keep receipts for any cleaning or repairs your kid paid for.

This isn't about being combative. It's about turning "I swear it was like that" into "Here's the video from move-in day."

Utilities and ``small" monthly fees that aren't small when they stack

Rent is the main bill, but it often comes with satellites. A listing might say "$1,500/month" and skip the part where the monthly total is closer to $1,750 once the orbiting charges settle in.

Common monthly add-ons:

- *Electric/gas/water/sewer/trash:* sometimes included, often not.
- *Internet:* sometimes an exclusive provider, sometimes tenant choice.
- *Parking:* paid spot, garage fee, or street permit.
- *Pet rent:* a monthly charge *in addition* to a pet deposit.
- *Amenity fees:* package lockers, gyms, "community technology fees," pest control.
- *Renter's insurance:* usually cheap, but still a line item.

A sharp question for your kid to ask on a tour: *"What will my typical total monthly cost be, including all required fees and average utilities?"* If the person showing the unit can't answer, your kid can still ask for a fee sheet and average utility ranges. "We can't estimate" sometimes means "We'd rather you discover it later."

Furniture and household basics: the quiet budget killer

The first time a young adult lives without parents, they discover a weird truth: houses come with stuff, and that stuff costs money.

The obvious items are big-ticket: bed, dresser, couch. The sneaky items are the "how is this $200?" purchases: shower curtain rod, cleaning supplies, spice rack, trash bags, hangers, a mop, a plunger (the most underrated $12 purchase on earth), and a doormat that makes the place feel like home.

A smart approach is to separate "needs to function" from "nice to have." A first apartment can be sparse without being sad.

Your kid doesn't need a fully curated Pinterest living room. They need:

- A safe place to sleep.
- A way to eat without bleeding money on takeout.
- Basic cleaning and maintenance gear to keep the deposit intact.

One of the best parent gifts isn't a fancy couch. It's helping them build a "household starter list" that prevents frantic, expensive convenience purchases at 10 p.m.

Rental scams: the landlord who's ``out of town" and the deal that's too good

When a young adult is anxious to move, scammers lean on the urgency like a crowbar. The classic script is painfully effective:

- The rent is below market "because they just want a good tenant."
- The landlord is traveling, deployed, or dealing with a family emergency.
- They can't show the unit right now, but they'll mail keys after payment.
- They push for cash-like payment methods: wire transfer, gift cards, crypto, or a payment app with no protection.

A second, newer flavor is less dramatic: a fake ad that looks real, with photos scraped from legitimate listings. The scammer collects application fees, "holding deposits," and personal information like Social Security numbers—which can create a second crisis later.

Teach your kid one simple line that can save them thousands: *Never pay to "hold" a place you haven't seen, and never hand over sensitive information until you've verified who you're dealing with.*

Verification doesn't require paranoia; it requires routine:

- Tour the unit (or at least confirm access through a legitimate management office).
- Compare the rent to similar listings nearby. A huge discount is a flashing warning light.
- Research the property manager/company name independently (not through links the "landlord" texts).
- Avoid irreversible payment methods for deposits and fees.

If your kid feels embarrassed needing to be cautious, remind them: scams don't work because victims are stupid. They work because the scammer is running a tight psychological play—urgency, scarcity, and the promise of relief.

The move-out plan starts on move-in day

The best renters aren't the ones with fancy furniture. They're the ones who make the end easy:

- They pay rent on time automatically.
- They report maintenance issues early (small leaks become big bills).
- They document the unit's condition.

- They keep a small "house buffer" in savings so a broken tire doesn't become a late rent spiral.

A late fee can feel like a slap, but financially it's worse: late payments can start a chain reaction of overdrafts, credit damage, and fractured roommate relationships. The antidote is boring: autopay rent if possible, and keep a cushion.

Actionable takeaway: the ``Before You Sign" script

Before your kid signs a lease, have them answer these questions in writing (notes app is fine). If they can't answer one, they don't have enough information yet.

1. *Total cash needed to move in:* application fees + deposits + first month + any required last month + utility deposits.
2. *Total monthly cost:* rent + required fees + realistic utilities + parking + pet rent + renter's insurance.
3. *Lease break terms:* what it costs to leave early, how much notice is required, and whether they owe rent until re-rented.
4. *Roommate liability:* whether each person is responsible for the full rent, and how rent will be paid.
5. *Deposit protection plan:* move-in photos/video completed, inspection checklist saved, and maintenance requests documented in writing.

If they do those five things, renting stops being a leap of faith and becomes a controlled decision. The apartment might still be imperfect—most first places are—but the finances won't be a surprise attack.

8.3 Transportation Costs and Decisions

A car ad is a magic trick performed with numbers. It doesn't sell you a 3,200-pound machine that drinks gasoline and slowly dissolves into repair bills. It sells you *a monthly payment*. Four tidy digits in a box. A promise that adulthood can be financed like a streaming subscription.

Then real life happens. Tires wear out. Insurance renews. Gas jumps. A warning light shows up with the emotional tone of a horror-movie violin. And your kid realizes the payment was never the whole cost—it was just the part that fit on a billboard.

The easiest way to protect a young adult from an expensive car mistake is to teach one phrase and insist they use it out loud:

> *"What's the total cost of ownership?"*

Total cost of ownership (TCO) is the real monthly cost of having a car in your life, not just buying it. It includes the costs you expect (loan payment) and the costs you'd rather not think about (everything else). When your kid can estimate TCO, they stop shopping for a car and start shopping for a *transportation plan*.

The car-as-asset myth (and the more useful truth)

A car is technically an asset because you can sell it. But for most households, most of the time, a car behaves like a liability: it pulls money out of your bank account predictably and then unpredictably. It's a tool that depreciates—a fancy word for "becomes worth less over time."

There are exceptions. Classic cars, rare models, collector markets. Those aren't transportation decisions; those are hobby investments. For a young adult trying to pay rent on time and build a life, the healthy assumption is simple:

> *Your car is a machine you rent from the universe. You pay for the privilege in monthly bills and occasional surprises.*

That mindset does something powerful: it makes "reliable" feel attractive. It makes "cheap to insure" feel cool. It makes "boring sedan" start to look like financial competence on wheels.

Total cost of ownership: the iceberg you can't afford to ignore

The monthly payment is the part of car ownership you can see. The rest is underwater, heavy, and absolutely capable of sinking a budget.

If that drawing feels dramatic, good. Transportation is one of the biggest slices of most household budgets. The car is often the most expensive object a young adult will buy besides education—and unlike education, it tends to go down in value.

The five big TCO categories (and the questions your kid should ask)

1) Depreciation: the cost nobody mails you a bill for

Depreciation is the financial equivalent of air leaking out of a balloon. Quiet, constant, and real whether you look at it or not.

If your kid buys a $22,000 car and can sell it for $17,000 three years later, that $5,000 is a cost of ownership. It doesn't show up as a monthly charge, which is exactly why people underestimate it. If they finance a car and later trade it in, depreciation is often what creates the gut-punch moment of being "upside down" (owing more than the car is worth).

A useful parent line: *"Every mile you drive has a tiny price tag."* Gas is one tag. Depreciation is another.

2) Insurance: the bill that punishes optimism

Auto insurance is a contract: you pay a premium so the insurer pays certain costs if something bad happens. Young drivers—especially teens and early-20s—often pay painful premiums because insurers price risk with cold math, not with your kid's good intentions.

The key teaching point isn't the vocabulary; it's the behavior: *insurance cost is part of the car choice.* Two cars with the same price can have very different insurance premiums. A sporty model with higher theft rates can cost far more to insure than a sensible compact sedan.

When your kid is car-shopping, "What's the insurance on that?" should be asked the same way as "What's the monthly payment?" Ideally, they get a quote *before* buying.

3) Fuel: the daily drip that becomes a monthly flood

Fuel is the easiest cost to understand and the easiest to ignore because it's paid in small chunks. A $45 fill-up doesn't feel like much until it happens four times in a month.

A quick way to estimate monthly fuel cost:

$$\text{Monthly miles} \div \text{MPG} \times \text{Price per gallon}$$

If your kid drives 900 miles a month, gets 27 MPG, and gas averages $3.50:

$$900 \div 27 \approx 33.3 \text{ gallons} \quad \Rightarrow \quad 33.3 \times 3.50 \approx \$117$$

That's the friendly month. Now add longer commutes, weekend drives, price spikes, or a car that gets 18 MPG, and the "small" category becomes real.

4) Maintenance and repairs: the difference between ``owning" and ``operating"

Maintenance is predictable: oil changes, filters, tire rotations, brakes, tires. Repairs are the plot twists: alternator, transmission issues, electrical gremlins, the mysterious rattle that turns into a four-digit invoice.

A young adult doesn't need to know what a serpentine belt is. They need a plan for what happens when the car demands money in a month they didn't plan for.

A helpful framing: *car ownership requires a "car fund."* Even $50–$100 a month set aside can turn a future repair from a crisis into an annoyance.

5) Taxes, registration, parking, and the ``life context" costs

Depending on where your kid lives, they may face registration fees, inspection fees, local taxes, tolls, parking permits, garage fees, and tickets. These don't feel like "car costs" until they show up. City life can make parking alone rival a car payment. Suburban life can make miles and fuel the larger monster.

This is why the best transportation decision is contextual. The "right" car in a rural town can be the wrong car in a dense city. The cost isn't just the vehicle; it's the environment you operate it in.

Case study: Payment thinking vs. TCO thinking

Sofia is 24 and just moved into her first apartment with a stable job. She wants a "nice" car because she's tired of driving the family hand-me-down. She sees a newer compact SUV for $28,000. The dealer offers a payment of $489/month.

That payment feels like the entire decision. Sofia does the mental math: "I can handle $489." She buys it.

Now compare two ways of thinking.

Sofia's original budget (payment thinking)

Line item	*Monthly*
Car payment	$489
Gas	$80
Total	*$569*

This is how people get in trouble: the visible costs look manageable, so the purchase feels "responsible."

Sofia's real budget (TCO thinking)

Let's use realistic-but-simple estimates:

- *Insurance:* $190/month (young driver, newer SUV, full coverage)
- *Gas:* $140/month (commute + errands)
- *Maintenance/repairs sinking fund:* $75/month (averaged)
- *Registration/taxes:* $20/month (averaged)
- *Depreciation:* say $200/month (varies widely, but it's real)

Now the same car looks like:

Line item	*Monthly*
Car payment	$489
Insurance	$190
Fuel	$140
Maintenance/repairs fund	$75
Registration/taxes (averaged)	$20
Depreciation (estimated)	$200
True monthly cost (TCO estimate)	*$1,114*

Sofia didn't buy a $489/month car. She bought an $1,114/month transportation plan. If her take-home pay is $3,400/month, this car eats about a third of it before rent, groceries, and life.

Nothing about this is meant to shame Sofia. This is how the marketplace is designed: the payment is easy to quote; the true cost takes work to compute. The win is teaching your kid to do the work.

Why used and reliable often wins (even when it's not glamorous)

If you want a sentence that can save a young adult five figures, it's this:

> *"A car's job is to be boring. If it's exciting, it's probably expensive."*

Buying used can reduce depreciation dramatically, because the steepest drop in value often happens early in a car's life. A reliable used sedan can also be cheaper to insure than a newer vehicle, and cheaper to repair than a complex luxury model.

But "used" isn't automatically "cheap." A cheap car with expensive problems is not cheap. The target is *reliable used*—the kind of car that starts every morning and doesn't demand a $1,200 repair because a sensor decided to retire.

Two practical tactics that keep "used" from turning into a nightmare:

- *Pay for a pre-purchase inspection.* A mechanic's inspection fee can be the best money your kid spends all year.
- *Avoid stretching for features that increase complexity.* Fancy tech packages and high-performance options can raise repair costs and insurance.

Leases, loans, and ``cash": how to talk about the three purchase paths

Young adults tend to treat the method of payment like a personality test: "I'm a lease person," "I'm a cash person," "I'd never finance." The better framing is: each path shifts *risk* and *cash flow*.

Financing (a loan)

A loan spreads the cost over time. It can be sensible when the car is reliable, the terms are reasonable, and the payment leaves room for the rest of life. The danger is financing too much car—or rolling old debt into a new loan so the payment looks small while the total obligation grows.

Leasing

A lease can lower the payment and put your kid in a newer car with a warranty, but it often includes mileage limits and wear-and-tear charges. It's less "owning" and more "renting a car long-term," which is fine if it fits the lifestyle and the math. It's not fine if it's used to pretend a more expensive car is affordable.

Paying cash

Cash avoids interest, which is great. But it can drain savings, which is dangerous. A young adult with $6,000 in the bank who spends $5,500 on a car may feel proud until the first medical bill, job gap, or repair shows up. Cash is only "safer" if it still leaves a buffer.

A simple rule that respects real life: *no matter how they buy the car, they still need a car fund.* The car doesn't care about the financing method when it needs tires.

The adult skill: matching transportation to the rest of life

A good transportation decision is one that makes the rest of your kid's goals easier, not harder. The right car:

- gets them to work reliably (income protection),
- doesn't force them to skip savings (future protection),
- doesn't turn every surprise into debt (stress protection).

That's why it can be smart to drive an older, paid-off car a little longer while building a cash cushion and credit. It's why "I can afford the payment" is the beginning of the conversation, not the end.

Actionable takeaway: a two-number test before your kid buys

Before your kid commits to a car, have them compute two numbers and write them down:

1. *Estimated TCO per month:* payment (or savings goal if paying cash) + insurance quote + fuel estimate + $50–$100 maintenance fund + registration/taxes averaged. Add a depreciation estimate if they're comparing new vs. used.
2. *TCO as a percentage of monthly take-home pay:* TCO ÷ net monthly income.

If the percentage makes them wince, listen to the wince. That's their nervous system doing math their optimism didn't want to do. The point of a car is to move their life forward; if it pins them down financially, it's not transportation—it's a weight.

8.4 The Safety Net: Insurance 101

Insurance is the most resented adult bill because it's a payment for a problem you hope never to have. You send money out every month, and if things go well, you get nothing back except the right to sleep at night. That can feel like paying a gym membership for a gym you never visit.

But here's the honest framing that makes insurance click for young adults: insurance doesn't exist to make you richer. It exists to keep one bad day from turning into a five-year financial injury.

A broken arm is painful. A broken arm *plus* a stack of medical bills you can't pay is life-shaping. A fender-bender is annoying. A fender-bender without liability coverage can become wages garnished, a license suspended, and a credit report that follows you into your next apartment application. Insurance is the safety net under the tightrope. You can be a skilled walker and still appreciate the net.

The core trade: premium vs. deductible (and why it feels backwards)

The two words that make people's brains short-circuit are *premium* and *deductible.*

- *Premium:* what you pay to keep the policy active (usually monthly).
- *Deductible:* what you pay out of pocket *when something happens* before the insurance starts paying (or starts paying more).

The relationship between them is a seesaw. If you choose a higher deductible, the premium often goes down. If you choose a lower deductible, the premium usually goes up.

This feels backwards to young adults because it sounds like: "Pay more later, pay less now." It can sound like a trick. It isn't. You're choosing how much risk you can personally absorb.

A higher deductible is basically saying, *"I can handle a bigger hit if something goes wrong, so I don't need to pay as much every month for the insurer to absorb small hits."*

A lower deductible says, *"I want fewer surprises. I'd rather pay more each month to reduce the size of a bad-day bill."*

Neither choice is automatically "smart." The smart choice depends on your kid's cash buffer and how stable their life is. A brand-new graduate with $400 in savings should be cautious about choosing a sky-high deductible just to shave $20 off a monthly premium. Cheap insurance that can't be used is a false bargain.

A simple test for choosing a deductible: can you pay it tomorrow?

Here's the practical parenting question that cuts through the noise:

> *If the bad thing happened tomorrow, could you pay the deductible without going into high-interest debt?*

If the answer is no, the deductible is too high for their current reality. That's not a character flaw; it's a cash-flow fact. The "right" deductible is often the one your kid can pay from savings, not from panic.

This is why the emergency fund and insurance are best friends. Insurance limits the size of catastrophic hits; savings handles the deductible and the small annoyances that insurance doesn't cover.

Insurance is a contract: teach your kid to read the one page that matters

Insurance policies can be thick and intimidating, which is why many people never read them. Fortunately, there's a shortcut: for many types of insurance, the most useful document is the *declarations page* (often called the "dec page").

The dec page is the policy summary. It typically shows:

- Who and what is covered
- The policy period (start/end dates)
- Coverage limits (the maximum the insurer will pay in different categories)
- Deductibles
- Premium amount

If your kid reads one insurance document in their life, make it the dec page. It's the difference between "I have insurance" and "I have *this* insurance."

A small but powerful exercise: have them pull up their auto or renters policy and circle the limits and deductibles. Then ask, "What's the most you'd have to pay out of pocket in a bad scenario?" You're teaching them to connect paperwork to consequences.

The "Big Three" for young adults: health, auto, renters

Young adulthood has a particular risk profile: income is rising but still fragile; savings are often thin; one surprise can force debt; and a single missed step can snowball into credit damage. That's why three policies are especially important.

Health insurance: not about checkups—about preventing bankruptcy

Many young adults skip health insurance because they feel invincible. They're healthy, they exercise sometimes, they "don't go to the doctor." That logic misunderstands the point of insurance.

Health insurance is less about routine care and more about putting a ceiling on financial disaster. A broken bone, appendicitis, a surprise ER visit, or an illness that requires ongoing treatment can create bills that dwarf what a young adult can handle.

Here's one term that matters more than almost any other: the *out-of-pocket maximum* (or out-of-pocket limit). This is the most you pay in a plan year for covered in-network services. Once you hit it, the plan generally pays 100% of covered benefits for the rest of that year. (Premiums don't count toward this number, and neither do non-covered services or out-of-network care.)

When your kid looks at a health plan, they tend to fixate on the monthly premium. That's understandable: it hits the budget every month. But the out-of-pocket maximum is the "how bad can this get?" number. Both matter. Premium is the rent. Out-of-pocket max is the fire escape.

Case study: Luis and the surprise out-of-pocket maximum

Luis is 23 and chooses the cheapest premium plan offered through his employer because he wants more take-home pay. He barely looks at the details. Six months later, he lands in the ER with severe abdominal pain and needs surgery plus follow-up care. The medical bills are large, but the bigger shock is how much he personally owes before insurance takes over.

Luis's plan has:

- A low premium
- A high deductible
- A high out-of-pocket maximum

He isn't "wrong" for choosing a low premium when money is tight. The mistake is choosing it without a plan to handle the deductible and out-of-pocket exposure. The right move would have been to pair the plan with a dedicated medical buffer (even $25–$50 per paycheck) or to choose a different plan if the risk was unmanageable.

The teaching point isn't fear. It's realism: health insurance is the policy that protects your kid's future income from being swallowed by one unexpected diagnosis.

Auto insurance: legally required, financially essential

Auto insurance is where a lot of young adults get tripped up because they assume insurance is about their own car. The most important part is often not the car at all. It's *liability*.

Liability coverage pays for injuries and damages you cause to other people. If your kid rear-ends a new SUV and it turns into a multi-vehicle accident with injuries, the costs can go far beyond the value of their own car. Without liability coverage (or with limits that are too low), your kid is exposed. That exposure can include lawsuits, wage garnishment, and years of financial cleanup.

Most auto policies also include categories like:

- *Uninsured/underinsured motorist:* protects your kid if someone hits them and doesn't have enough insurance.
- *Collision:* covers damage to your kid's car from an accident (often with a deductible).
- *Comprehensive:* covers non-collision damage (theft, hail, falling objects, etc.).

A young adult with an older, paid-off car may choose to drop collision/comprehensive to reduce the premium, accepting they'll self-insure the car's value. That can be a rational choice *if* they understand the trade and can handle replacing the car if it's totaled. What's usually non-negotiable is liability coverage at levels that won't collapse their life.

A practical habit: any time your kid is quoted a premium, have them also read the liability limits on the quote. Price without limits is like buying a parachute based on color.

Renters insurance: cheap, underrated, and often the easiest "yes"

Renters insurance is the policy that people skip because their stuff doesn't feel valuable. Then a pipe bursts, a kitchen fire starts in a neighboring unit, or a laptop gets stolen, and suddenly "my stuff isn't worth much" turns into "why am I rebuilding my life with a credit card?"

Renters insurance usually covers:

- *Personal property:* your kid's belongings (up to a limit)
- *Liability:* if your kid is found responsible for injury or property damage (and often legal defense)
- *Additional living expenses:* if the unit becomes uninhabitable due to a covered loss

Another key detail: the landlord's insurance generally covers the building structure, not the tenant's belongings. Your kid can be a perfect tenant and still lose their stuff to a fire that started two doors down.

Renters insurance is often surprisingly affordable, which is why it's a great first lesson in smart risk transfer. If your kid is allergic to "adulting," renters insurance is a gentle gateway drug.

The hidden risk young adults forget: liability is everywhere

Young adults often think risk is about their own things: my car, my health, my laptop. But the scariest costs usually come from harm to *others*.

- A car accident injures someone.
- A guest trips in the apartment and gets hurt.
- A dog bites a neighbor.

These are liability events. Liability is the category that can turn a person with a modest income into someone facing a terrifying bill. That's why liability coverage is the grown-up heart of insurance, even when it's boring to talk about.

How to shop without drowning: the ``same limits, then compare price" rule

Insurance shopping becomes a mess when your kid compares apples to dragonfruit. One quote is cheap because it has low limits and a high deductible. Another is higher because it covers more and costs less in a claim.

A clean way to compare:

1. Pick the coverage limits and deductibles you want.
2. Ask multiple insurers to quote *that same setup.*
3. Compare premiums and service reputation.

This turns shopping into a real comparison instead of a "which number is lowest" trap.

When to say no to extra insurance (because not every policy is a hero)

Once your kid understands insurance, they'll notice it's sold everywhere: at checkout counters, during ticket purchases, when renting a car, when buying electronics. Much of this is designed to monetize anxiety.

A good rule: say yes to insurance that protects against *catastrophic* loss you can't afford, and be skeptical of insurance that protects against minor inconveniences you could cover with savings.

The goal isn't to be anti-insurance. It's to be intentional. A young adult who buys every add-on is still anxious, just poorer.

Actionable takeaway: the 15-minute insurance audit

Give your kid one short assignment that makes insurance real:

1. Pull the declarations page (or summary) for auto and renters insurance (or request it).
2. Write down four numbers on paper:
    - Monthly premium
    - Deductible
    - Liability limit
    - Out-of-pocket maximum (for health insurance, if applicable)
3. Answer one question: *If a bad thing happened this weekend, what's the maximum cash I'd need quickly?*

If that number is larger than their current savings, the next step isn't panic. It's a plan: adjust deductibles, shop for better pricing with the same limits, and build a small buffer so the safety net can actually catch them. Insurance works best when it's paired with a little cash and a clear understanding of what it does—and what it absolutely does not do.

8.5 The Independence Checklist

Independence is often pictured as a dramatic moment: the keys, the goodbye hug, the first night in a place that's all yours. In real life, financial independence is quieter. It's the ability to absorb a normal amount of chaos without making an emergency phone call. It's not "never needing help." It's "not being one minor surprise away from a crash landing."

A checklist sounds unromantic, which is exactly why it works. Romance is for movies. Rent is due on Tuesdays. The goal here is to help your kid build a small set of systems that keep their life from turning into a constant series of financial fires.

Think of this as a pre-flight check. Not because life is an airplane, but because the same logic applies: most disasters come from ignoring simple warning lights.

Start with the one number that decides everything: the monthly burn rate

Young adults often know their hourly wage, or their salary, and still don't know the most important number in their financial life: how much it costs them to exist for one month.

Burn rate isn't a budgeting buzzword. It's a survival metric. It answers: *If income stopped tomorrow, how long could I keep my life intact?*

To calculate it, your kid needs two pieces:

- *Net monthly income* (take-home pay, not gross).
- *Fixed monthly commitments* plus realistic variable spending.

What makes this hard for new adults is that their first few months are messy. They're buying household basics, paying deposits, and dealing with a schedule that doesn't feel stable yet. That's fine. The goal isn't perfect precision. The goal is to replace "I think I'm fine" with "I know what fine costs."

A strong first step is to choose a baseline month (not move-in month, not vacation month) and track every expense for 30 days. Not forever. Just long enough to get honest.

The *minimum viable budget*: a plan simple enough to actually follow

Many young adults hear "budget" and imagine spreadsheets, guilt, and never having fun again. The emotional resistance is real, and it's often rational: they've seen budgets used as punishment.

A better way to frame it is: *a budget is a spending plan that prevents accidental self-sabotage.* It's a system for making sure the essentials get paid before the money evaporates into a dozen small purchases that felt harmless in the moment.

For a first-time independent adult, the budget does not need to be detailed. It needs to be *durable*. A minimum viable budget usually has:

- *Bills that must be paid to avoid consequences:* rent, utilities, insurance, minimum debt payments.

- *Daily life:* groceries, transportation, basic personal needs.
- *A small "future" line:* savings, even if it's tiny.
- *A guilt-free fun line:* because a plan that allows no joy is a plan that gets abandoned.

If your kid can keep those four categories running, they're not just budgeting; they're building the habit of financial self-command.

Autopay is a superpower (with one important warning)

Late fees are a tax on disorganization. For a young adult, late fees also tend to arrive in clusters: one late bill triggers an overdraft, which triggers another late bill, which triggers a stress spiral.

Autopay breaks that chain. The goal is to automate the bills that can hurt them:

- Rent (if possible)
- Utilities
- Car insurance
- Minimum debt payments

The warning is equally important: autopay without oversight is how you bleed money quietly. Teach your kid to do one monthly ritual: open the bank app and scan transactions for five minutes. Automation should reduce mental load, not eliminate awareness.

A good compromise is to set bills on autopay and set a calendar reminder two days before the big ones hit, so the account balance is ready. Autopay is not a replacement for cash-flow management; it's a tool that rewards it.

The emergency buffer: one month changes everything

The most dramatic improvement your kid can make in their financial life is not a perfect budget or an investing strategy. It's a small cash buffer.

Even $500 can turn a car repair from "credit card crisis" into "annoying Tuesday." One month of expenses can turn job turbulence from panic into planning.

The phrase "emergency fund" can sound intimidating, like it has to be $10,000 in a high-yield account or it doesn't count. For a young adult, a better target is a ladder:

- *Level 1:* $500 buffer (stops the bleeding from small surprises).
- *Level 2:* one month of bare-bones expenses (rent, utilities, food, transportation, insurance).
- *Level 3:* two to three months (especially important for variable income or unstable work hours).

The psychological shift here is enormous. A buffer doesn't just pay bills; it buys time. Time is the ingredient that turns bad decisions into tolerable ones: "I need cash today" is when people take predatory loans, sell things in a panic, or move in with the wrong roommate.

A concrete case study: Nina's first solo month

Nina is 21 and moving out for the first time. She has a steady job, no debt, and feels ready. She's also never fully run her own household. The first month is a stress test.

Here are her numbers:

- Net pay: $2,850/month
- Rent: $1,250
- Utilities + internet: $210
- Car insurance: $165
- Gas: $130
- Groceries: $320
- Phone: $55
- "Life" (toiletries, laundry, random stuff): $120
- Fun: $150

Her baseline burn rate is about $2,400. That leaves roughly $450.

Month one, two things happen:

- She needs two new tires: $420.
- Her friend gets engaged and she travels for one weekend: $260.

If Nina has no buffer, she ends the month using a credit card and telling herself she'll "catch up later." That "later" often turns into a recurring pattern, not because she's irresponsible, but because her system wasn't designed for normal surprises.

If Nina has even a $500 buffer, she can pay for the tires, still make rent, and then choose whether the trip fits her budget rather than letting social pressure choose for her. The buffer turns her into the decision-maker again.

The paperwork trap: the small admin tasks that quietly cost real money

New adults often lose money not through big splurges, but through unfinished paperwork:

- Wrong tax withholding because a W-4 was rushed or never updated
- Payroll deductions they didn't understand (or benefits they meant to enroll in but missed)
- Missing a deadline for employer benefits, leaving them uninsured for months

The checklist approach here is simple: your kid should know what's being withheld from each paycheck and why. They should confirm the pay rate, hours, and net deposit match expectations. This isn't about becoming an expert; it's about catching errors early while they're easy to fix.

A practical habit: save the first pay stub from any new job and the first pay stub after any raise or benefit change. Those are the moments when mistakes happen.

Credit: not a score, a reputation

Young adults often think credit is a number that appears when you "start adulting." In reality, credit is a reputation system:

it's a record of whether you pay back what you promised, on time, over time.

That record matters in more places than your kid expects:

- Renting an apartment
- Setting up utilities
- Buying a car with financing
- Sometimes even job applications (depending on role and local rules)

If your kid understands credit as reputation, the behaviors become obvious: pay on time, keep balances manageable, avoid unnecessary debt, and check the record occasionally for errors.

The grown-up move: read the credit report, not just the score

A credit score is a snapshot. A credit report is the underlying file. If something is wrong, the score can wobble, but the report is where you find the cause.

Many people don't realize they can access their credit reports for free on an ongoing basis through the official channel (AnnualCreditReport.com). The key habit is not obsessive monitoring; it's periodic review. A young adult who checks a few times a year is more likely to catch errors, fraud, or old accounts they forgot existed.

A credit report check is like looking at your medical chart: you're not doing it because you expect disaster, you're doing it because you don't want surprises to be the first time you see a problem.

Banking setup: the two-account system that prevents accidental broke

If your kid uses one checking account for everything, their financial life will always feel slightly chaotic. The money for rent and the money for tacos live in the same place, and whoever shows up first wins.

A simple system that works for many young adults:

- *Bills checking:* rent, utilities, insurance, minimum debt payments. Income flows in, bills flow out. Keep a cushion.
- *Spending checking:* groceries, gas, fun, everyday life.

This isn't about hiding money from yourself. It's about making it harder to accidentally spend rent. When bills money is separated, your kid can enjoy spending money without that low-level dread that they're stealing from future obligations.

If two checking accounts feel like too much, a variation is one checking plus one savings bucket that's labeled "Bills Cushion" or "Do Not Touch." Labels help. So does friction.

The ``who pays when I'm sick?" conversation (aka: stop pretending independence is solo)

Independence doesn't mean your kid never needs help. It means the default plan isn't "call someone." It also means everyone knows the rules before stress hits.

A powerful family exercise is to define the boundaries in calm weather:

- If your kid has a medical bill, what happens? Do they call you before it goes to collections? Do they pay a minimum? Do you help if they're insured but can't cover the deductible?
- If they lose a job, can they come home temporarily? For how long, and under what conditions?
- If the car breaks down, is there an emergency loan option within the family, or is that on them?

This isn't about control. It's about avoiding the worst version of "help": last-minute crisis help that arrives with resentment and confusion. Clear expectations make support cleaner and more dignified.

The Independence Checklist (printable mindset, real-life version)

Here's a system check that fits on one page in your kid's brain. Each item has a reason attached—because adults don't do chores; they do *cost avoidance*.

Money flow

- Net pay known (per paycheck and per month).
- Bills and due dates listed in one place.
- Autopay set for the bills that can cause cascading damage.
- A monthly "scan" ritual scheduled (5 minutes, not a therapy session).

Buffer

- $500 starter buffer (minimum).
- One month of bare-bones expenses identified as the next target.
- A plan for where the money sits (savings account, separate bucket, not in the spending stream).

Credit and identity

- Credit report accessed and reviewed at least once.
- Any errors disputed and documented.
- Accounts and passwords secured (because fraud is a financial event, not just a tech event).

Housing and transportation

- Lease obligations understood (especially early termination and roommate liability).
- Car costs estimated as TCO, not just payment.
- Insurance in force (health, auto, renters).

If your kid can confidently check most of these boxes, they're not just "moving out." They're building the systems that keep them out.

Actionable takeaway: the Sunday night drill

Pick one Sunday night this week. Set a timer for 30 minutes. No shame, no lectures, no "you should have done this earlier." Just a drill.

1. Write down net monthly income.
2. List all bills with due dates and minimum amounts.
3. Move $25 into a dedicated buffer account (even if it feels small).
4. Pull one credit report and scan for anything unfamiliar.
5. Choose one thing to automate before next payday.

That's it. Thirty minutes. A small transfer. One report. One automation. Independence doesn't arrive as a mood. It arrives as a few boring systems that quietly keep your kid's life from becoming expensive.

8.6 Bouncing Back: Resilience and Recovery

The most expensive part of a money mistake is often not the fee. It's the fog.

A young adult overdrafts for $32, misses a payment by one day, or realizes they've been paying for three subscriptions they don't remember signing up for. The numbers aren't fun, but they're survivable. What turns a fixable problem into a long-running mess is the emotional cocktail that shows up next: embarrassment, avoidance, and a weird belief that "good with money" people never mess up.

They do. They just recover faster, with less drama, because they have a script.

Resilience is not a personality trait. It's a sequence of actions you practice until it feels normal: notice quickly, contain the damage, clean it up, and change the system so it doesn't keep happening. Shame is the enemy because it delays the first step, and delay is where fees multiply.

The first rule of recovery: speed beats elegance

When something goes wrong, many young adults try to solve it perfectly. They research for hours, draft the ideal email, wait until they "understand everything." Meanwhile, the clock is running:

- Late fees stack.
- Interest accrues.
- Accounts get reported as delinquent.
- Collections become possible.

The recovery skill is not "know everything." It's "act while the problem is still small."

A useful mantra for your kid: *today is always the cheapest day to fix it.*

The OOPS method: a simple loop that turns mistakes into a system upgrade

Here's a recovery loop that works for most common mistakes. The name matters because it's hard to remember wisdom when you're panicking.

OOPS: Own it. Organize a fix. Pay it. Systematize.

It's not cute. It's a way to get your kid out of a shame spiral and into action.

Own it: identify what happened without self-insults

Owning it doesn't mean groveling. It means describing the situation like a grown-up:

- "My checking account went negative after autopay hit."
- "I missed my credit card payment and got charged a late fee."
- "I forgot to transfer money for rent, and the payment bounced."

Notice what's missing: "I'm so stupid." "I'm terrible with money." "I always do this." Those are identity statements, and identity statements keep people stuck.

A helpful parent response when your kid confesses a mistake is not "How could you?" It's, "Okay. What's the exact timeline?" Calm is contagious. So is panic.

Organize a fix: freeze the situation before you solve it

This step is about damage control. The goal is to stop the bleeding.

For most common money mistakes, organizing a fix means:

- Find the relevant account, bill, or transaction.
- Identify what is due, to whom, and by when.
- Determine whether anything is about to happen automatically (another autopay, another fee, another overdraft).

This is where many young adults avoid looking because they're afraid of what they'll see. Teach them a trick: *don't look for the full solution; look for the next action.*

The next action might be as small as "turn off autopay for 48 hours" or "move $60 from savings to checking so nothing else bounces." Small actions create air.

Pay it: the fastest path back to clean ground

Once the situation is contained, pay what needs to be paid. Not because the company deserves it, but because your kid deserves a clean slate.

If they can't pay the full amount, the goal shifts to *communication*:

- Call the company.
- Explain the situation briefly.

- Ask what options exist: payment plan, due-date change, fee waiver, hardship program.

A lot of young adults think calling will be humiliating. It's usually not. Customer service representatives talk to people in financial chaos all day. A calm, responsible caller who wants to fix the issue is often the easiest person they'll speak with.

A key coaching point: your kid should ask directly for the thing they want.

> *"I missed the payment by accident and I've paid it today. Can you waive the late fee as a one-time courtesy?"*

Sometimes the answer is no. Often the answer is yes, especially if it's the first offense. But a fee is guaranteed if they don't ask.

Systematize: the part that turns ``mistake" into ``lesson"

Systematizing is what separates adults who repeat the same problems from adults who improve. It means asking, *What in my system made this likely?* Then changing the system.

Common system upgrades include:

- Moving due dates to align with paydays.
- Setting low-balance alerts on checking accounts.
- Keeping a small "checking buffer" that never gets spent.
- Using two accounts: one for bills, one for spending.
- Turning on autopay for minimum payments (and paying more manually).

The goal isn't perfection. The goal is fewer emergencies.

Case study: The overdraft that became a monthly habit (until it didn't)

Caleb is 20, working part-time and taking classes. He keeps $50–$150 in his checking account most of the time and uses his debit card for everything. One week, his phone bill and a streaming subscription hit within hours of each other. He buys groceries in between. His account goes negative. Overdraft fee: $34.

Caleb is angry, but he's also embarrassed. He doesn't tell anyone. He thinks, *I'll just be more careful.*

Two weeks later, it happens again. Another $34. Now it feels personal, like the bank is punishing him. He starts avoiding his banking app because looking makes him feel worse. That avoidance causes another overdraft. Within two months, Caleb has paid over $100 in fees for the privilege of being stressed.

Here's how the OOPS loop fixes it:

- *Own it:* "My account balance runs too close to zero for autopay to be safe."
- *Organize a fix:* He turns on low-balance alerts at $100 and $50. He checks his upcoming autopays.
- *Pay it:* He transfers $120 from savings, bringing checking positive and preventing more fees.
- *Systematize:* He creates a "do not touch" checking cushion of $200 and moves subscriptions to a credit card with autopay set to minimum payment (so they can't cause a

checking overdraft). He also changes the phone bill due date to two days after payday.

The most important change wasn't the $200 cushion. It was the shift from "I need to be better" to "my system needs guardrails." Caleb didn't become a different person. He built a safer environment for the person he already is.

Missed payments: triage based on what gets ugly fastest

Not all missed payments are equal. If your kid misses one, the first move is triage: fix the thing with the biggest consequences first.

In general, the "gets ugly fast" category includes:

- *Rent:* risks fees, eviction processes, and future rental history issues.
- *Car payment/insurance:* risks repossession or being uninsured.
- *Utilities:* risks shutoff fees and disruption.
- *Credit cards and loans:* risks late fees, interest, and credit reporting.

The fastest adult response is to contact the biller *before* things escalate. Companies can be surprisingly flexible when you communicate early and pay something.

A key detail to teach: credit reporting often doesn't happen immediately after one late day. Late fees can hit quickly, but many lenders don't report to credit bureaus until an account is 30 days past due. That means there's a window where fast action can prevent long-term credit damage. The window is not an invitation to procrastinate; it's a reason to sprint.

Credit report errors: how to dispute without losing your mind

Sometimes your kid does everything right and their credit report still shows something wrong: an account that isn't theirs, a balance that's incorrect, a late payment that shouldn't exist.

The process for disputes is boring, but it's powerful. The key points:

- Dispute with the credit reporting company (the bureau) *and* with the furnisher (the company that provided the information, like a bank or lender).
- Be specific: what's wrong, what should be true, and what documentation supports it.
- Keep records: dates, screenshots, confirmation numbers.

Your kid may fear that disputing is like arguing with a wall. It can be frustrating, but it's worth doing because errors can raise borrowing costs and affect housing approvals.

One more detail that makes this less mystical: credit reporting companies can reject disputes they consider "frivolous" or irrelevant if the consumer provides too little information. That's why specificity matters. "This is wrong" is weak. "This account is not mine; I have never lived at this address; attached is my ID and proof of address" is stronger.

Identity theft and scams: treat it like a fire drill, not a moral failing

If your kid gets scammed or has their identity stolen, the emotional load can be brutal. People feel foolish, violated, and ashamed. That shame can delay the actions that limit damage.

Here's the reframe: identity theft is not a reflection of intelligence. It's the result of exposure plus opportunity. Your kid can be cautious and still get hit. The goal is fast containment.

A practical starting point is the government's guided recovery tool at IdentityTheft.gov, which helps victims generate an Identity Theft Report and build a step-by-step plan. The value isn't just the paperwork; it's the sequence. When your brain is overloaded, sequence is oxygen.

If the scam involved renting (a common scenario), the response often needs two tracks: money recovery (reporting, payment disputes) and identity protection (credit reports, fraud alerts, account changes). Your kid doesn't have to do it all in one night. They do have to start.

The ``repair your mindset" part: replace shame with curiosity

Some kids grew up in households where money mistakes were treated like moral failures. Others grew up where money wasn't discussed at all, so mistakes feel like evidence of incompetence.

A resilient mindset uses a different language:

- Instead of "I'm bad with money," say "My system has a gap."

- Instead of "I always mess up," say "This failure mode is predictable."
- Instead of "I can't look," say "Looking is how I make it smaller."

Curiosity is the opposite of shame. Curiosity asks, "What set this up?" Shame says, "Don't touch it." One builds adults. The other builds avoidance.

Actionable takeaway: the 24-hour recovery protocol

When your kid makes a money mistake, the most useful gift you can offer is a clear script for the first 24 hours. Here's a protocol that fits most situations:

1. *Name the mistake in one sentence.* ("My checking overdrafted because two autopays hit.")
2. *Stop the bleeding.* Move money, pause autopay if needed, prevent a second fee.
3. *Contact the right party.* Bank, lender, landlord, utility—whoever can waive a fee, reverse a charge, or set a plan.
4. *Pay the minimum to return to good standing.* Full amount if possible; otherwise, a plan with dates.
5. *Install one guardrail immediately.* Low-balance alert, due-date change, autopay for minimum payments, or a separate bills account.
6. *Write down what happened.* Two sentences in a notes app. Not to punish yourself—to build a personal playbook.

The point of a protocol is not to eliminate mistakes. It's to make mistakes small, brief, and educational instead of big, chronic, and identity-shaking. A young adult who can recover quickly doesn't just protect their money; they protect their confidence, which is the resource they'll need for every other grown-up decision.

www.ingramcontent.com/pod-product-compliance
Lightning Source LLC
LaVergne TN
LVHW052023200926
848481LV00005B/1565
* 9 7 9 8 8 9 6 6 5 2 9 6 0 *